COMMITTEE ON PROJECT FORMULATION
FOR IRRIGATION AND DRAINAGE SYSTEMS

The purpose of the Committee on Project Formulation For Irrigation and Drainage Systems is to carry out technical activities relating to project formulation, including consideration of: 1) the objectives and need for projects, 2) physical resources available, 3) engineering, economic, social, environmental, legal, and financial aspects, and 4) interrelationships with other water uses.

The Committee held its first meeting in September 1972. In 1974, the Committee concluded that there was a need for a manual on the Principles of Project Formulation For Irrigation and Drainage Projects and initiated discussions of how to proceed. In April 1977, the Committee Chairman presented a detailed outline of the manual at a technical session of an ASCE conference in Dallas. At the National Convention in Chicago in October 1978, drafts of some of the manual chapters were presented as technical papers by the Committee member authors. At the July 1979 I&D Specialty Conference in Albuquerque, papers on the remaining chapters were presented at one of the technical sessions by other Committee member authors. The individual chapters were revised, considering the comments received at the technical sessions; and in April 1980, the first complete draft of the manual was compiled. That draft was reviewed by the Committee members and by other individuals with expertise in project formulation. Based on those comments, preparation of a revised edition was undertaken. At their March, 1981 meeting, the I&D Executive Committee decided that it would be preferable to publish the document as a special report rather than a manual, which would have required prior publication in the I&D Journal. In August 1981, a revised edition of the report was completed and submitted to Committee members and to the I&D Executive Committee for final review. Additional revisions were made in 1982 and with the approval of the Executive Committee, the document was submitted for publication as an ASCE special report.

It is intended that after the report has been in circulation for a few years and subjected to peer group review, it will be revised and published as an ASCE Manual.

ACKNOWLEDGMENTS

The authors gratefully acknowledge the direct and indirect contributions to this report by a number of agencies and individuals. In particular, the cooperation of The World Bank is acknowledged for permitting publication in Chapter 5 of some of its methodology and examples of economic and financial analyses. Also, acknowledgements are due the U.S. Water Resources Council, Bureau of Reclamation, and Soil Conservation Service for their development of some of the methodology and procedures referred to in this report. This report would not have been possible if professionals involved with irrigation and drainage projects throughout the world had not taken the time to prepare the numerous documents which are referred to.

Special thanks and appreciation go to Mrs. Cynthia G. Erivez who typed drafts and final copy of the report and assisted with organization and editorial review.

Comments On This Report
Should Be Addressed To:

American Society of Civil Engineers
Irrigation and Drainage Division
345 East 47th Street
New York, New York 10017

J. MAJUMDAR

PRINCIPLES of PROJECT FORMULATION for IRRIGATION and DRAINAGE PROJECTS

A report prepared by the Technical Committee on Project Formulation For Irrigation and Drainage Systems of the Irrigation and Drainage Division of the American Society of Civil Engineers

COMMITTEE CONTROL MEMBERS

W. Martin Roche*—Chairman
Otto J. Helweg—Vice Chairman
Michael R. Stansbury—Secretary
Karl R. Klingelhofer*+—Executive Committee Contact

COMMITTEE CORRESPONDING MEMBERS

George R. Baumli*+   David B. Palmer*+
Frederick L. Hotes*+   Richard R. Schaefer*
James N. Krider*+   Lonnie D. Schardt

+Past Chairman, included Dan Lawrence and Rowland Fife
*Authored portion of Report

Edited by George R. Baumli

Published by the
American Society of Civil Engineers
345 East 47th Street
New York, New York 10017

The material presented in this publication has been prepared in accordance with generally recognized engineering principles and practices, and is for general information only. This information should not be used without first securing competent advice with respect to its suitability for any general or specific application.

The contents of this publication are not intended to be and should not be construed to be a standard of the American Society of Civil Engineers (ASCE) and are not intended for use as a reference in purchase specifications, contracts, regulations, statutes, or any other legal document.

No reference made in this publication to any specific method, product, process, or service constitutes or implies an endorsement, recommendation, or warranty thereof by ASCE.

ASCE makes no representation or warranty of any kind, whether express or implied, concerning the accuracy, completeness, suitability or utility of any information, apparatus, product, or process discussed in this publication, and assumes no liability therefor.

Anyone utilizing this information assumes all liability arising from such use, including but not limited to infringement of any patent or patents.

FOREWORD

Large increases in food production are necessary for the world's increasing population. Much can be obtained by improving operations and management of existing irrigation systems and by placing more lands under irrigation. Irrigation and drainage projects are one of the more effective means of closing the gap between world demand for food and fiber and supply.

Irrigation and drainage projects should be formulated to accomplish their intended purpose with full consideration of physical, economic, social, and environmental factors. It is to this end that this report "Principles of Project Formulation for Irrigation and Drainage Projects" has been prepared. The objective of the report is to provide guidelines to ASCE members and others who are engaged in the formulation or irrigation and drainage projects. There are numerous common elements in the formulation of irrigation and drainage projects regardless of their size, scope and location. Some projects have been successful, some have not. This report sets forth the generally accepted and proven principles of project formulation and provides a guide and checklist for the planning and review of irrigation and drainage projects.

Project formulation involves a series of steps starting with determination of objectives by the decision makers, identification and definition of problems and needs, evaluation of available resources, development of alternative means of resolving problems and meeting the needs, evaluation of the alternatives and selection and implementation of the recommended plan. It is an orderly and systematic process which permits the interested public and decision makers to become aware of the assumptions made, data used, rationale and methodology employed, alternatives considered, cost, benefits, impacts and consequences of the alternatives and throughout the process to play a role in the decision making process.

The report was prepared by the Committee on Project Formulation For Irrigation and Drainage Systems of the Irrigation and Drainage Division. Helpful comments were received on draft sections of the report when they were presented as papers at technical sessions of the Society's conferences. Accuracy, clarity, and usefulness of the report were enhanced by the constructive suggestions of a panel of expert reviewers.

Hopefully, the first edition of this report will serve as a foundation on which to build a more comprehensive and complete reference manual. Your comments are invited.

W. Martin Roche

W. Martin Roche, Chairman
Committee on Project Formulation
For Irrigation and Drainage Systems
August, 1982

TABLE OF CONTENTS

TABLE OF CONTENTS

TABLE OF CONTENTS

TABLES

TABLE OF CONTENTS

FIGURES

CHAPTER 1. INTRODUCTION

It is the purpose of this report to provide guidance to those individuals who have the responsibility to formulate irrigation and drainage
projects and to review formulation studies prepared by others. It is
recognized that regardless of the unique nature of a specific project,
the formulation process has many common elements and the experience
gained from one project is applicable to another project to a significant degree. The report, on the one hand, attempts to set forth
generally accepted principles of project formulation and, on the
other, to provide a check list for the planning and review of irrigation and drainage studies.

The general procedures set forth herein should be viewed in relation
to the particular physical, temporal, and economic setting of the
project under consideration.

Irrigation is man's application of water to land for growing plants.
Measures to reduce soil moisture content are called drainage. Irrigation and drainage are complementary processes to maintain soil
moisture required for optimum plant growth.

An irrigation project generally consists of a storage dam and reservoir or a diversion dam or pumping plant on a stream, a system for
conveying the regulated water to the farmer's headgate, an on-farm
distribution system, and a system for collecting the unused water and
returning it to the stream system for subsequent use by downstream
diverters. Groundwater can be used in lieu of or in conjunction with
surface water. Drainage facilities can consist of subterranean tile
collector pipes, and deep surface ditches to collect the excess water.
Pumps are frequently required for both the delivery of irrigation
water and for the disposal of drainage water.

The varying degree of success of existing projects indicates the need
for thorough and realistic formulation studies. History can provide
us examples of projects which are successful, others which have fallen
short of achieving the goals which were envisioned, and others which
were failures. Although, not always well documented, the knowledge of
success and failure is available to project planners and it only
remains for such knowledge to be diligently sought out and applied to
the project at hand. However, as resources become scarcer and the
demands for them become greater and more diverse, it becomes increasingly necessary to improve our ability to formulate effective,
efficient, and acceptable projects. It is toward this end that this
report is presented.

The report is organized to discuss each of the steps of the project
formulation process.

1

Steps of the Project Formulation Process

The six steps of the project formulation process which are intended to be carried through in increasing detail for each level of study are:

1. Determine objectives of the decision makers.
2. Determine need for project.
3. Inventory available resources.
4. Develop alternative plans.
5. Evaluate and compare alternative plans.
6. Select and implement plan.

Step 1. Determine Objectives of the Decision Makers

Irrigation and drainage projects can be formulated in different ways to meet different objectives. The manner in which projects are formulated, constructed, and operated is determined by the decision makers. The success of the project depends upon the reasonableness of the objectives and availability of resources to formulate, construct, and operate and maintain the project. Objectives of irrigation and drainage projects may include improving living conditions by increasing employment, increasing personal income, and improving social conditions; improving national economic efficiency; improving foreign exchange balances; and improving the distribution of population.

Step 2. Determine Need for Project

The need for an irrigation and drainage project can best be determined in the larger context of the need for food and fibre. The economic need for a specific irrigation and drainage project is usually expressed in terms of market demand for a product or service provided by the project. In a multiple-purpose irrigation project, one or more of the following purposes may be included: drainage, flood control, hydroelectric power, municipal and industrial water supply, navigation, and recreation.

Step 3. Inventory Available Resources

The success of an irrigation and drainage project depends on a thorough inventory and analysis of available physical, financial, and human and institutional resources. An assessment of their quantity, quality and constraints is a necessary precondition to developing alternatives. The physical resources which should be evaluated include climate, land, soils, water, plants and animals, energy, transportation, aquaculture, man-made facilities, archeological and historical resources and aesthetics. In addition to the physical resources, it is important to consider financial resources, which are the tangible assets needed to plan, design, construct, and operate and maintain a project to fulfill its intended objectives.

Step 4. Develop Alternative Plans

The development of alternative plans is necessary to insure that the most favorable solutions to the problems are considered. The alternative of improving an existing irrigation project as well as achieving

2

the objectives through non-structural means should be considered along with the alternatives involving construction of new facilities. There usually will be competition or conflict between objectives, as the achievement of one may reduce the achievement of the others. Other factors which contribute to the need for formulating alternative plans include limited resources, technical planning constraints, acceptability, legal, institutional and administrative constraints, and implementable stratagies. The initial list of alternatives should be developed without screening or ranking based on cost or other constraints. Even possibly nonviable proposals should be included if they have significant public interest and support. It is necessary to document that these plans were considered and justification given for not selecting them for further analysis. The economic costs and benefits and environmental and social effects should be developed on a comparative basis for each of the alternatives.

Step 5. Evaluate Alternative Plans

The plans must be evaluated to determine how well they meet the objectives. The differences among the alternative plans must be analyzed to show trade-offs among the specified components of the objectives. The beneficial and adverse effects of each alternative must be evaluated in terms of effectiveness, completeness, efficiency, and acceptability.

Steps 4 and 5 are iterative and may need to be repeated several times.

Economic, financial, and environmental and social analyses are critical elements in plan evaluation. Economic and financial analyses are closely related in the decision making framework, however, they have important differences. Economic tests are made to estimate total return, productivity or profitability to society as a whole from the viewpoint of a needed investment. Financial analyses are made to measure the ability of beneficiaries to meet their financial obligations, and when appropriate, to estimate returns to equity capital, labor, and management. Environmental and social analyses are made to determine beneficial and adverse effects of projects on the environment and society, to provide a basis for selection of the plan which minimizes adverse effects, and to provide a basis for mitigation of those adverse environmental effects which cannot be avoided. The above factors are then collectively analyzed to determine the best plan or if a project can be justified at all.

Step 6. Select Plan

The recommended plan should be one that meets the objectives, is publicly acceptable, provides maximum flexibility in meeting needs, and minimizes adverse environmental effects. Given such criteria, there should be no more economical means of accomplishing the purposes of the project and net benefits should be maximized. Total economic, environmental, and social benefits should exceed total economic, environmental, and social costs and each separable purpose should provide benefits at least equal to its cost unless there is a stated exclusion. In some cases, it may not be practical to meet all these requirements, but they are the goals for which to strive.

The six steps of the project formulation process are carried through in increasing detail for each of the following levels of study.

Levels of Study

Planning is an iterative process. That is, the process is carried out in a number of stages, each with more detailed data and analysis than the last. The stages or levels of study are continued until the desired level of definition is achieved, consistent with the established objectives and available resources. In some cases because of time and budget constraints, it may not be possible to complete each level of study, including the report, before moving into the next level.

The decision to study the project in more detail, to go to a more intense level of study, is based on the results of the previous level. If, for example, based on a reconnaissance evaluation, a proposed project shows promise it normally would be studied in more detail; if not, the project normally would be dropped from consideration.

Following is a brief description of various levels of study usually followed in the planning of irrigation and drainage projects.

Level I. Reconnaissance - (Sometimes called Preliminary Project Investigation, or Pre-Feasibility Study)

This is the first level of study. It consists of a preliminary appraisal of water and land resources problems and possible alternative solutions to determine whether further investigation and expenditure of funds for a more detailed study is warranted on the basis of existing conditions. Field work, research, and office studies are held to the minimum necessary to meet these objectives. Supporting economic data are preliminary in nature, but sufficient to enable identification of the most favorable solutions. Preliminary benefit-cost and financial analyses and environmental and social evaluations are included. If the results are favorable, a decision is made to carry the more promising alternatives to the second level of study, namely:

Level II. Feasibility - (Sometimes called Survey)

As the name implies, the purpose of this level of investigation is to determine engineering, economic and financial feasibility of a proposed project and to define environmental and social effects. A feasibility report, which is based on detailed field and office studies, includes definition of all project features and operations for a number of alternatives in sufficient detail to define approximate project costs, accomplishments and environmental and social effects. Economic and financial studies are carried to sufficient detail to identify project beneficiaries, and to determine the overall investment requirements and to identify sources of financing. The feasibility report and accompanying environmental documentation generally serves as the basis for the decision to commit the necessary resources to implement the project. Once the project is authorized or approved, evaluations shift to the third level of study, namely:

4

Level III.   Implementation Plan - (Sometimes called Definite Plan,
Work Plan, General Design Memorandum or Advance Planning)

The purpose of this level of study is to define the features of that
selected plan in sufficient detail to further determine specific costs
and accomplishments. In addition, many years or decades may have
elapsed since the feasibility study was completed, and modifications
are required to bring the project study up to date. Adjustments are
made as necessary in the financial aspects to insure funding. The
results of the implementation plan provide the basis for design,
construction, and operation of the project.

Public Participation

At each step of the plan formulation process and for each level of
study, it is important to consider public participation. Public
participation in the formulation of irrigation and drainage projects
is emphasized more in developed countries than in undeveloped countries.
In developing countries, public participation may have only limited
applications.

The primary objective of public participation in the formulation of
plans for irrigation and drainage projects is to identify all alternatives and possible environmental and social effects. Public participation promotes citizen trust in the fairness and objectivity of the
planning process. A public participation program should insure that
government officials (local, state, regional, and national), influential
and opinion leaders, and the general public are informed and involved
in the decision making process. The program should provide a process
by which the public can participate in a visible manner in the steps
that lead to decisions which directly or indirectly affect those
interested in the study, and should be initiated early in the planning
process and incorporated in activities throughout the entire study.

Public participation involves two-way communication to:

o   Keep the public fully informed regarding the status and
    progress of studies and the results and implications of
    planning activities.

o   Obtain from involved interest groups their opinions and
    perceptions of problems, issues, concerns, and needs; their
    preferences regarding resource use and development of alter-
    native managerial strategies, and any other information and
    assistance relevant to the plan formulation process.

o   Use public input to influence the project formulation process
    and to assure that a full range of alternatives is considered.

Considerable effort must be devoted to the public participation effort
in planning studies, particularly where serious conflicts in public
desires and values are likely to arise. Early identification of such
situations will allow time for the planning organization to more
effectively design and conduct the public participation program most

likely to be successful in dealing with sensitive value-conflict situations. Care should be taken to avoid overwhelming the public with well-intended but unspecific efforts. The objective of each action should be clearly defined. Public participation meetings should essentially be scheduled in relation to the six steps of the project formulation process. However, if it is a highly visible project or there are significant conflicts involved, additional meetings should be scheduled.

There will often be competing demands for water resource use. To the extent possible, plans should be formulated that are responsive to the problems, needs, and concerns of the public involved in the planning process. To facilitate comparisons and trade-offs among alternative plans and comparisons of beneficial and adverse effects measured in non-monetary terms with beneficial and adverse effects measured in monetary terms, alternative plans should be formulated which emphasize each objective. Consideration should also be given to formulating non-structural plans for components such as flood or erosion control, and to describing future conditions without any plan of development. When considering alternative plans which reflect major trade-offs between conflicting objectives, the addition of complementary measures to serve several objectives may considerably enhance the plans.

There will be uncertainty as to what the public consensus may be regarding trade-offs and, indeed, decisions cannot be reached until the range of trade-offs is made available to the public and feedback obtained. Therefore, a variety of alternative plans may need to be developed initially which appear to represent the preferences of the various public interests. During subsequent iterations, the alternatives can be refined, and those which lack significant public support can be eliminated. The number of alternatives which are to be carried through to the end of the planning process is a function of both the diversity of public and professional expressions and the characteristics of the possible plans formulated to meet planning objectives.

CHAPTER 2. OBJECTIVES OF DECISION MAKERS AND NEED FOR PROJECT

The basic objective of an irrigation and drainage project is to provide food and fibre. It follows that the need for such a project can best be determined in the larger context of the need for food and fibre.

The demand for food is determined by the number of people in the world, by their income, and by their individual preferences and customs. These factors interact with supply to determine market prices and the actual amount of food that the world's population will consume.

Large increases in food production are necessary for the world's increasing population. Much can be obtained by improving operations and management of existing irrigation systems or by placing more lands under irrigation. The opportunity for bringing new lands under irrigation is the greatest in developing countries; opportunity for improvements in irrigation systems exist in already irrigated areas in both developed and developing countries.

Not only are there great opportunities for improving operation of existing irrigation projects, but much can be learned about how to plan, design, construct, and operate successful new irrigation and drainage projects by studying problems of existing projects. Some examples of problems with existing irrigation systems follow.

About one-third of the irrigated land in the world is adversely affected by salinity and high water tables. In India, about 6 million hectares have gone completely out of agricultural production because of salt and sodium accumulations in the soil. Iran has about 24 million salt-affected hectares. In West Pakistan, 40,000 hectares per year are going out of production because of salinization. In the 17 western states of the U.S.A., about 28 percent of the irrigated lands suffer depressed crop yields because of soil salinity buildup.

These types of problems can result from a number of factors, acting singly or in combination: irrigation with high salinity waters, high salinity soils, poor drainage, inefficient water use and water management, and excessive application of fertilizers.

The framework within which problems of existing irrigation systems are solved and the objectives to be achieved are in many ways similar to the framework involved and the objectives in the formulation of new irrigation projects.

Objectives of Decision Makers

Irrigation and drainage projects can be formulated in different ways to meet different objectives. The manner in which projects are

7

formulated, constructed, and operated is determined by the decision makers. The success of a project depends upon the reasonableness of the objectives and commitments to the resources required to formulate, construct, and operate the project. It is assumed that the decision makers set the objectives and control the resources. Control may be exercised as a consequence of resource ownership, custodial responsibility, or political leadership.

Objectives of irrigation and drainage projects, in addition to providing food, may include:

o   Improving living conditions by:
     - increasing employment
     - increasing personal income
     - improving social conditions
o   Improving national economic efficiency
o   Improving foreign exchange
o   Improving distribution of population
o   Protecting and enhancing the natural environment

Improving Living Conditions

Increasing Employment

Benefits may accrue to a region if irrigation development results in increased employment or greater utilization of the underemployed. Benefits may be short term (duration of construction) and long term. In most cases, an irrigation project will result in increased long-term employment, including those involved in project operation and maintenance, and those directly and indirectly involved in farm operations, and those involved with processing and marketing of farm products.

In assessing the potential effects of an irrigation project on employment, account should be taken of all project purposes. Flood control, for example, may result in more intensively used lands; recreation may not only provide an outlet for the recreationist, but create new businesses and services catering to the recreationist.

Unless the aggregate supply of consumer goods and services available to the workers and farmers increases in proportion to the changes in incomes, additional employment may only trigger a rise in the price level leaving only owners of land and capital as the beneficiaries.

Increasing Personal Income

Increases in personal income can result from construction of irrigation and drainage projects. The change from dry farming to irrigation can increase crop production and permit the growing of different higher profit crops. Income benefits consist of the increased value of goods and services and are reflected in increased economic activity and additional net income induced by the project.

Impacts upon income distribution can result from project development. A development in a rural, one-industry setting will have a more sig-

8

nificant impact than a comparable development in a large, diverse economic setting. Additional water supply can result in utilization of undeveloped resources and stimulate new economic activity.

Income distribution effects resulting from water projects will be influenced by: 1) uses to which the water will be put; 2) nature of the existing economy; 3) land ownership, in some cases; 4) where the water will be used; and 5) the cost. A payment capacity analysis can be used to assess additional incomes to farm workers and incomes generated in support industries can be derived from salaries and wages and any change in employment levels. The relative share of income in the agricultural sector can then be compared to other sectors of the economy.

A change in water supply accompanied by a change in land ownership adds another dimension to the analyses. More, but smaller farms, suggest a redistribution of income. It does not, however, necessarily mean an increase in total farm income or a redistribution within related industries. The distribution effects are apt to be more pronounced within farming than between farming, other industries, or the community. It is a matter subject to measurement on a case-by-case basis.

Whether increased productivity will be shared by the bulk of the population or concentrated among a few will depend in part upon other economic circumstances that surround the financing, operation, and disposition of project services. Careful integration of engineering and agricultural activities with economic policies may be required if the project benefits are to accrue to those most in need.

In the case of government-owned projects, the distribution of income can be determined by administrative action. Where projects are developed by private enterprise, distribution of income is determined by market processes.

Improving Social Conditions

One objective of an irrigation and drainage project can be to induce social change among rural populations. As part of a project, farmers are introduced to new farm practices in new setttings and are usually ready to adapt to modernized techniques. However, without training this may be slow in happening because the difficulties of adjusting to strange environments and new agricultural methods are substantial. Traditional social practices can be disrupted as a result of migration or new farming practices.

Part of an irrigation project's benefits can be in the form of services, such as: safe water for domestic use, sanitation systems, housing, electric power, communication, and health transportation facilities. However, in many areas, most of these benefits must be provided by social programs.

When an irrigation project is established in a semiarid area, the prospect of permanent water and prosperity inevitably attracts people

from other areas with different backgrounds and traditions. For
example, some may be nomadic and unaccustomed to living and working in
an area with others. Special considerations, including training,
patience, and understanding of different customs, are necessary in
order to transform a new irrigated area and its new workers into a
productive permanent agricultural development.

Other factors may be of great importance in influencing social res-
ponses. One is a change in work pattern, resulting from almost continuous activity through the year when two to four crops are grown on
the same ground under perennial irrigation. A reduction in leisure
time ard a change in timing of different activities through the day,
month, and year alter the whole pattern of life compared, say, with
accompanying grazing stock. Farmers' cooperative effort in such
matters as water use, mechanization, marketing, and financial assistance becomes crucial to project success.

Improving National Economic Efficiency

The formulation of plans for an irrigation and drainage project can be
directed towards meeting current and future national needs, problems,
and opportunities. National economic development components normally
involve values to users of goods and services resulting from project
purposes.

Large irrigation projects can have substantial impact on the national
economy through: 1) source of funds, 2) changes in expenditures for
goods; and 3) anticipated inflationary or deflationary effects. Once
the project is in operation, other multiplier effects stem from the
increased demand for agricultural inputs and from foods and fibers
that are stored, transported, or processed.

Improving Foreign Exchange Balances

A project objective can be to contribute to exports to help improve
foreign exchange balances. Where market forces are dominant within
the economy and incomes are evenly distributed, production of export
crops may be more profitable than production for the domestic market.
While it is possible that the bulk of workers may benefit from export
markets, such benefits would come only if the resulting demand for
labor leads ultimately to higher real wages, a result that could
depend on appropriate governmental actions as well as market forces.

Improving Distribution of Population

A new project is likely to stimulate the growth of towns or cities as
well as increasing the density of rural population. Moreover, the new
residents are likely to be relatively young adults who will establish
households and raise families. With enhanced food supplies, increased
employment opportunities, reduced flood hazards, and better water,
sanitation, and electrical facilities, powerful forces for rapid population growth may come into play. These forces are likely to be
augmented by the addition of families dislocated by the project who
received new houses and promises of employment. The increased density
of population can create new opportunities for the people but also

10

public health and safety problems, all of which should be considered in evaluating project impact.

Where a project is initiated in an area characterized by nomadic or extensive pastoral activity, other benefits may be taken into account such as increase in economic activity that follows from provision of seed, fertilizer, insecticides, herbicides, equipment, fuels and labor. To achieve additional economic activity, the range of needs associated with creation of new cities, towns, and villages must be considered.

Protecting and Enhancing the Natural Environment

The decision maker should consider the environmental effects of a project proposal and incorporate, to the extent practical, features to enhance positive environmental effects and to minimize any adverse environmental effects.

Examples of positive environmental effects that can be created by irrigation and drainage projects:

o   Safe and dependable supply of water for irrigation cf crops and drinking
o   Food, employment, and improved living conditions
o   Electricity for homes and industries
o   Reduced flood damage
o   Reduced soil erosion
o   Improved riparian habitats for fish and wildlife
o   Fishing, swimming, and boating areas

Examples of adverse environmental effects that can be created by irrigation and drainage projects :

o   Reservoir inundation of fertile lands, settlements, or historic and archeological sites
o   Destruction of habitat for fish and wildlife
o   Degradation of downstream water quality
o   Increased water logging, salinization, and erosion of farm lands
o   Reduction of downstream flows of water, sediment, and nutrients through estuaries to the seas, thereby damaging aquatic life.

Understanding the environmental opportunities and complications arising from irrigation and drainage projects in large measure depends upon the communication among the people concerned. The decision maker, project manager, administrator, engineer, economist, environmentalist and farmer--each makes decisions that affect the total project. The overall effectiveness of the project depends upon the amount of coordination between these participants.

The Decision to Proceed

The decision to go ahead with a new irrigation and drainage project depends upon 1) whether it is the most economical way of adding to

11

food and fiber resources compared with the use of more fertilizer, pesticides, better seeds, or improved cultivation practices on land already under cultivation -- dryland and irrigated or transfer of production to another region; 2) whether an existing project can be improved more readily than a new project can be established; 3) whether, given a limited supply of capital, social conditions are better off with an irrigation project rather than more schools, housing, health services, roads or other forms of capital investment; 4) whether it is the most appropriate type of land use for the land capability; and 5) whatever is compatible with environmental and social goals.

If the project appears to be justified on the basis of broad policy and budgetary considerations, project formulation can proceed to more specific matters including: 1) determination of the economic need for the project; 2) determination of project purposes; 3) evaluation of the physical, financial, and human resources available for the project; 4) formulation of alternative plans; and 5) evaluation of benefits and costs.

Need for Irrigation Project

The economic need for a specific project is usually expressed in terms of demand for a product or service provided by the project. In the case of an irrigation and drainage project, the need can be expressed in terms of market demand for agricultural crops or food. Market demand, or effective demand, is an expression of the quantity of a product or service that can be sold at given prices and terms under given market conditions.

Schedules of expected market demand for project products and services generally provide an upper limit to the physical dimensions of a project. However, availability of water subject to regulation and availability of irrigable land also can be constraints. Determination of the amount of acreage to be irrigated is based on the need for food and fiber and the capacity of the region under consideration to meet all or a part of that need. The amount of water needed for an irrigation project depends primarily on the amount of irrigable land, soils, climate, growing season, type and distribution of crops, and type of irrigation management.

The demand for crops in a given study area is based on past and present cropping patterns; land characteristics, type and quantity of soils; climate; project share of the region's production; income returns and the availability and cost of water. Projecting the historical relationship between a study area's cropping patterns and that of the region may serve as the basis for a first approximation of cropping patterns and irrigated acreage. The projections may be modified by available land, quality of soils, and climatic constraints, and availability and cost of water. In those situations where the study area may be undeveloped, historical relationships are not relevant. In such cases, the cropping patterns of areas having similar characteristics may be used as a guide.

12

Necessary crop rotations, land committed to perennial crops, and long-standing cultural practices are all given consideration in arriving at projected crop patterns for an area. Profitability of double cropping will also have an affect on the farmer's cropping decision and on total irrigated acreage and should be considered.

Projections of irrigated acreage should reflect the extent that urbanization may remove land from production or affect the value of land to the point that the farmer turns to higher income crops or ceases to operate.

Once projected acreages are established, irrigation water needs are determined. The amount of water required depends on the consumptive use of the crops to be grown, soil leaching requirements, and conveyance losses, with allowances for the portion of the crop water requirement provided by precipitation. Requirements may vary by region, depending on soils and climate. Total irrigation water requirements are determined by multiplying the projected acreage by the unit applied water requirements.

In the case where farmers will be required to pay for their water, payment capacity or the ability to pay for agricultural water is determined. It is the maximum ability of the bulk of farmers in a specific area to pay annual average costs for water at their farm headgate, on a unit volume basis, over a specified repayment period. It is the difference between gross returns from the sale of crops and the cost of production, including an imputed cost for the farmer's own labor, management, and risk, except the cost of water.

Payment capacity analysis can be used to establish the upper limit of agricultural water user's market demand for project water, and is one of the elements used in demand analyses for financial feasibility studies. It is also used to determine the lower limit below which lands are not capable of producing sufficient returns to be considered economically suitable for irrigation. Payment capacity is discussed further in Chapter 5.

Benefits from irrigation water are computed as the increased net returns that would result from reduced production cost, production of higher profit crops, and increased acreage. Benefits are determined as the difference in net income with project compared to without project.

Need for Drainage

Irrigation and drainage are complementary processes to maintain soil moisture required for optimum plant growth. Measures to reduce soil moisture content are called drainage. Drainage features often are included in the initial development of an irrigation project, but in many cases, the drainage system is installed many years after the lands have been under irrigation. In most projects there are areas which ultimately will require drainage systems. Such drainage needs should be given early consideration and the costs thereof should be included in the project formulation.

Drainage prevents adverse effects of high water tables and waterlogging of the soil, provides for control of excess salinity and alkali accumulations in arid areas, and prevents soil erosion from excess wastewater. Drainage is essential whenever salinity is involved. Salt concentrations in the root zone must be maintained at tolerable levels by leaching. This involves the application of more water than required by the crop (within limits) and measures to drain the excess water.

The need for drainage is based on topography, soil conditions, type of crops grown, quality of water and other factors. The demand for drainage in a developed area is a function of net farm income. A demand exists if the lack of drainage threatens to make a farming operation uneconomic or if returns on an investment in drainage facilities exceed the costs. Farm budget analysis may be used to determine the effect of drainage facilities on income and whether the operation remains an economic unit. Crop patterns, yields, and farm costs are estimated on the basis of the drainage conditions expected to occur with and without the project.

Benefits from drainage projects are computed as the increased net returns resulting from reduced crop production cost or intensification benefits from increased production of current crops and increased acreages of new crops. The intensification benefits are computed as the difference in net income "with project" compared to "without project."

Increased acreages of new crops are evaluated as the efficiency gained in the project area compared to typical lands in the area.

Need for Other Project Services

A determination must be made of the general type of project to be pursued; whether a single purpose irrigation project will satisfy the objectives or whether a multiple purpose project is required, including one or more of the following purposes, flood control, hydroelectric power, municipal and industrial water supply, navigation and recreation.

Flood Control

Provision for flood control may be incorporated into irrigation and drainage projects. Flood protection for both agricultural and urban areas is important. However, the following discussion covers only the agricultural aspects since urban flood control is adequately covered in other literature.

The amount of reservoir storage dedicated to flood control depends upon the benefits from the flood control storage as compared with those from using that same amount of storage for conservation of irrigation water. Consideration should be given to alternative methods of reducing flood damage such as floodplain zoning and flood proofing.

The most fertile soils are usually located in floodplains and maximum agricultural productivity of such areas requires protection from damaging floods.

14

Agricultural flood control benefits usually are measured as the increased value of agricultural output or the reduced cost of maintaining a given level of output. The benefits include but are not limited to reductions in production costs, in associated costs, and in damage costs from floods, erosion, or sedimentation, the value of increased production of crops; and the economic efficiency of increasing the production of new crops in the project area. The benefit is measured by net agricultural income with the project compared to without the project.

Hydroelectric Power

Hydroelectric power is commonly included as a purpose in an irrigation project. The size and justification of the installation depending upon the available head, the amount of flow, and the demand for the power.

A hydroelectric plant can operate at a relatively low speed, compared to a steam generating plant, and can be quickly brought up to full output to meet changes in electric loads. The economic value of a hydroelectric plant can be increased by installing more capacity and utilizing the available amount of water flowing through the plant to meet the maximum utility system loads. For these reasons, within the limits of engineering feasibility and economic justification, and compatibility with the irrigation purpose, the installed capacity at each site should be the maximum capacity, with the associated energy, that can be absorbed by the power market.

Power is commonly sold at market value. The principle measure of market value at the project generating plant is the cost of producing and delivering to the market equivalent alternative power, with appropriate adjustment for the cost of transmitting and/or wheeling project power to the market.

Municipal and Industrial Water Supply

Municipal and industrial (M&I) water supply needs are usually small in comparison to irrigation water needs. However, in some cases provision for an adequate M&I water supply of adequate quantity and quality, as part of an irrigation project, may be necessary or at least desirable. M&I water needs usually are derived by forecasting population changes and multiplying those estimates by current average per capita water use factors. In areas where large water using industries are contemplated, water use for those industries can be estimated by applying use data for similar industries which relate water use to unit of product, employment, area of plant, or other parameters. Determination of M&I water needs should take into consideration water conservation measures. Benefits for M&I water supply are usually based on the cost of the most likely alternative.

Navigation

Navigation is usually handled separately, but can be incorporated into large irrigation and drainage projects. The estimate of demand for navigation facilities is based on projected relationships of water-borne commerce to area population, past and present cargo movements, industrial activity, limitations and opportunities afforded by project conditions, and competing means of transportation. The basic economic benefit of a navigation project is the reduction in costs required to transport commodities.

Recreation

Man is naturally attracted to water and creation of a new lake or waterway will result in recreation use. It can be planned for and incorporated as a project purpose. The demand for recreation is expressed in terms of user-days. These estimates are translated into specific requirements concerning service areas, types and volumes of probable use, facilities and services required, and need for access roads.

Potential recreation use at proposed water facilities is a function of population and per capita participation. A simplified approximation may be achieved by determining the total population in areas of origin; average per capita demand; and percent of participation based on travel time from population centers to project site.

Multiplying the population by the per capita demand results in the total demand (participation days) one could expect of that population. Actual participation in a recreational activity at any given site, however, is a function of travel time. Generally, the greater the time, the smaller the participation. The relationship may be expressed as: Participation days equals population times per capita demand times percent participation by travel time zone.

The demand for fishing and hunting as recreation activities may be determined in a manner similar to that described for recreation generally.

The principles for evaluating the physical, financial, and human resources available for the project, formulation of alternative plans, evaluation of costs and benefits and selection of the plan are discussed in succeeding chapters.

References

1. "A Hungry World: The Challenge to Agriculture," University of California Food Task Force, July 1974.

2. "Arid Land Irrigation in Developing Countries: Environmental Problems and Effects," E. Barton Worthington, Editor, February 1976.

3. "Contributions of Irrigation and Drainage to World Food Supply," ASCE Speciality Conference, August 1974.

4. "Economics Practices Manual Draft," California Department of Water Resources, April 1977.

5. "Guide to the Economic Evaluation of Irrigation Projects," Organization for Economic Co-Operation and Development, 1976.

6. "Irrigation, Drainage, and Salinity, An International Source Book," FAO/UNESCO, 1973.

7. "Irrigation Efficiencies in Producing Calories and Proteins: An Annotated Bibliography," California Water Resources Center, February 1975.

8. "Procedures for Evaluation of National Economic Development Benefits and Costs in Water Resources Planning," U.S. Water Resources Council, December 1979.

9. "Proposed Rules, Principles, Standards, and Procedures for Planning Water and Related Land Resources," U.S. Water Resources Council, April 1980.

CHAPTER 3. IDENTIFICATION AND DESCRIPTION OF RESOURCES

Physical Resources

Introduction

The physical resources evaluated should not be limited to those of direct importance to the irrigation and drainage project. A broader base is necessary to adequately evaluate potential environmental effects of the project. The following narrative covers the physical resource elements related to:

o Climate	o Energy Resources
o Land	o Transportation
o Soils	o Aquaculture
o Water	o Manmade Facilities
o Plants and Animals	o Archeological and Historical
o Air Quality	Resources
o Visual Quality	

First, a strategy for the evaluation must be developed based cn the project objectives and scope. An effort should be made to identify the more important resources, their interdependence, the size of the project area, and the degree of detail required for the investigation. These factors are all important in planning the evaluation of physical resources, and they should be documented in a written plan of study that will cover all elements of planning.

The plan of study usually covers three levels of planning: reconnaissance, feasibility, and implementation. Many elements of the physical resource investigation will be repeated in each level; the later ones being of increased intensity. The first level may be largely a reconnaissance survey, review of existing literature and data, and discussions with local people to make use of their knowledge. Duplication of efforts, "or of previous work," should be avoided. All activities and information gained in each phase should be documented in writing.

An interdisciplinary team for performing the investigation of physical resources is most important, especially in the reconnaissance phase. Specialists can recognize visual characteristics of an area during their first visit that may be critical in determining the scope of further investigations or the whole project. The assembled specialists should work as a team and not carry on to their separate conclusions without communication with other team members. There must be ample opportunity to alter the physical resources investigation or that of the total project at any time, based on the information gained.

Although it would take a large proposed project to justify the full range of specialists concerned with physical resources, the following fields should be covered. One person may be able to cover more than one field, but an investigation where only one or two people try to cover all areas probably will be inadequate. The specialists most often needed are:

o Planning Engineer	o Botanist
o I&D Engineer	o Biologist
o Water Quality Engineer	o Forester
o Soil Scientist	o Archeologist
o Geologist	o Environmentalist
o Agronomist	o Sociologist
o Hydrologist	o Economist

An evaluation system should be developed at the same time at the plan of study. A network analysis may be necessary for the systematic tracing of cause and effect sequences. There are natural cause and effect relationships and those that will be induced by the project. In the physical resources investigation, the former is of major concern, and the future without project condition is an important aspect.

Also, consideration should be given to the overall classification of lands in the project area. Considerable information on project goals must be available to effectively classify lands in a manner that will provide useful information for the planners. The primary factors in land classification are soils, topography, drainage, productive capacity, cost of production, and land development. The interdependence of economics, soil science, and engineering is inherent in the land classification process.

Land classification specifications must be developed for each project. The specifications used for one project should not be applied to another unless and until careful economic review indicates that it is possible to do so. Final land classification results will be no better than the competence of the professionals involved and the cooperation among them.

The following discussion covers each of the physical resource elements; the level of detail, procedures, or specialists involved are not covered.

Climate

Climate exerts important influence on the land. The characteristics of the soil, drainage, native vegetation, and crop adaptation are closely related to climate. Except for extremes, climate can be offset by intensive management, including frost control for example. A reasonably favorable climate greatly reduces the complexities and risks in project design and operation. The planners must know the characteristics and consequences of the climate in the project area. Important features are:

o Daily maximum and minimum temperatures during growing season
o Number of frost-free days
o Minimum temperatures during winter (winter kill)
o Wind velocity and direction
o Humidity, solar radiation, and evaporation
o Potential for hail, snow, and excessive wind
o Rainfall characteristics. Variations in annual, seasonal, and hourly rates are important. Also, the chemical content of the rainfall may be important.
o Air drainage for some types of cropping systems

The characteristics of climate are generally responsible for consideration of a drainage or irrigation project. Possibly, the major problem is that precipitation comes during the wrong time of year or that soil characteristics are not compatible with the amount received.

Climate may be the most difficult factor to evaluate since long-term records are needed and these may not be available. Short-term data should be collected and used with caution. A trained specialist may be able to translate short-term data into a reasonable prediction of the long term by correlating them with other observations (tree rings, etc).

Land

Land as referred to in this section includes topography, geology, existing use, and general surface characteristics.

A good map of the area is a requirement. Ultimately, a detailed topographic map of the project area will be needed for planning. Maps will be used for recording the information collected in the inventory of physical resources. Detailed maps as well as graphs and other tabular information on physical resources will be needed. More specifically, the following should be included in this evaluation:

o General topography. The general topography of the entire drainage area should be mapped to a point downstream where the effects of the project are expected to be negligible. Land slopes, elevations, stream channels, existing land use, general vegetative cover and, geometry of needed reservoir sites, should be shown. Manmade features such as settlements, roads, bridges, and railroads should also be shown.

o Detailed topography. The principal topographic character istics determining land suitability are degree of slope, relief, and position. Natural drainage characteristics such as source of excess water, adequacy of outlets, and potential for diversions should be recorded. A detailed topographic map including the location of pertinent manmade and physical features and vegetative cover will be needed for the project site. Elevations may be to an assumed datum or sea level, depending on the size of the project and its relationship to external features. A 1-foot contour map will ultimately be needed for the potential cropped area for most drainage and

21

irrigation projects. (A greater interval will be acceptable for sprinkler or drip irrigation systems.) The scale of the map is dependent on the features to be planned, but it will normally range from 1:1200 to 1:6000.

o Geology. It is important to know the origin and geologic framework of the entire drainage basin. Items of importance are:

    o Geologic and geochemical characteristics
    o Ground water aquifers and recharge areas
    o Ground water levels, springs, and seeps
    o Faults
    o Earthquake susceptibility
    o Possibility of land and snow slides
    o Mineral resources (coal and other economically important mineral deposits)
    o Caverneous rock areas
    o Potential of accessible formations for construction materials
    o Water holding capabilities of potential water storage sites
    o Stability of stream channels
    o Existing active wind and water erosion
    o Abandoned and active underground mines

Rock outcrops, existing open cuts, and well or mine logs all provide useful information to the geologist. Some exploratory drilling may be needed in the early stages of investigation, and considerable drilling will be needed before the investigation is complete. Complete logs of geologic materials should be kept, and appropriate samples collected for analysis.

Low altitude stereoscopic aerial photography, and magnetic and seismic surveys photogrammetric techniques, may fill some investigation needs. Vegetative cover and the level of detail needed will determine their appropriateness. Images from LANDSAT may also be useful in early stages of investigation.

Soils

The importance of suitable soils for a successful drainage or irrigation project cannot be overemphasized. To find out the limitations of the soils in the project area, investigations are necessary. These investigations will include hand and mechanical borings, open pits for the visual examination of the profile, and the collection of samples for analyses.

Basically, the soil will be evaluated for overall suitability and the ease of getting it in shape to be planted, short and long-term productivity, and the cost of maintaining the soil in a highly productive state. Factors that enter into this evaluation, some of which are interdependent are:

o Water holding capacity
o Infiltration and permeability
o Soil temperature fluctuations
o Natural drainage characteristics and ease of needed artificial
  drainage (ground water level, artesian pressures and anticipated fluctuations)
o Stoniness
o Susceptibility to flooding, water erosion, and wind erosion
o Natural fertility
o Physical and chemical characteristics of each soil layer in
  root zone (shrink-swell potential, mineral and organic con-
  stituents, bulk density, grain size distribution, pH, etc.)
o Soil depth suitable for plant growth
o Drainage characteristics of substrata soils

Soils in a project area generally are not uniform. Therefore, an
accurate soils map of the area is necessary for a complete evaluation
to show the location and extent of each kind of soil. The characteristics of each soil should be described for future planning activities.

Water

Water is essential to plant growth. Too much or too little can be
disastrous. However, water can be managed to get rid of the excess and
to make the best use of that available on-site or added from off-site.
Both the quantity and quality of water are important. The evaluation
must consider both surface and subsurface water sources and the rights
thereto. Existing lakes, ponds, marshes, springs, streams, and ground
water aquifers must be studied. Data needs will vary depending on
whether the project is primarily for irrigation or drainage. Quantity
is frequently given the most attention, but quality can prove to be a
deciding factor for the long-range success of a project.

Quantity and availability of water supply are of utmost importance for
irrigation. For irrigation, the potential amount of water supply that
can be made available from each reasonable source must be determined.
Evaluations must be on an annual, monthly, and, in cases, daily basis
for average and dry years. If surface storage is required to achieve
this potential, it must be so identified. Consideration should be
given to magnitude and frequency of floods. Ground water availability
is difficult to evaluate. If there are no existing wells in the area
that can provide proper data for evaluation, it will be necessary to
install test wells and run pumping tests. Drainage may be needed with
or without irrigation. The quantity to be disposed of and the availability of an outlet must be determined. Conjunctive use of surface
and ground water systems should be considered.

The quality of water affects plant growth, soil condition, the functioning of project structural features, and downstream uses of the
runoff. A chemical and physical analysis should be made of all water
related to project operation. Some of the water properties of concern
are silt, chemistry, pH, organic matter, seeds, spores, bacteria,

fertility, bio-chemical oxygen demand, coliforms, toxics, and salt content. The inventory should consider possible changes in quality at the same monitoring point from one time to another and the reason for identified changes.

Plants and Animals

In the "Land" section, it was recommended that existing land use and vegetative cover be inventoried in a general way. However, additional evaluations are needed to provide data and insights on the potential impacts on existing vegetation and animal life. The agronomist, botanist, and biologist can assist in this evaluation. Factors to consider and insights to be gained are:

o Possible adverse effects of a project on the habitat of any rare or endangered species and other "important" species.
o The diversity and population of existing fish and wildlife species in the project area.
o The growth, diversity, and production characteristics of existing plants.
o Relationship between plant and animal life in the project area and adjacent areas.
o Evidence of existing vegetation (timber).
o Quantity, quality, and seasonal variation of instream flows.

A desirable project objective is to increase the production of plant and/or animal life. There is much to be learned from the existing condition of this physical resource that will be of use to the perceptive planner.

If the area is already partially developed, the productive capability of domestic plants and animals in the present environment must also be assessed.

Air Quality

Although air quality is a basic environmental parameter, its importance in most drainage or irrigation evaluations is minor unless the project is in the vicinity of a source of pollution--an urban or industrial setting. Some air quality factors for consideration are pollen, smoke, dust, odor, and polluting gases such as carbon monoxide, sulphur dioxide, hydrogen sulfide, etc.

Energy Resources

Implementation of a new project or rehabilitation of an old project requires energy for construction, operation, and maintenance. The availability (quantity, quality, and characteristics) of energy resources should be part of the physical resources inventory. The project will have to be designed with consideration of the availability of energy, its costs, the appropriateness of its use, and the cultural and social characteristics of the area.

Transportation

Transportation facilities to, from, and within the project area must be
evaluated to determine:

o The ability to bring in needed equipment, materials, and
people for construction.
o The accessibility of the area to laborers, supplies, and
equipment during the farming operations.
o The ability to transport products to markets in a timely and
efficient manner.

Aquaculture

The features of an irrigation and drainage project may provide opportunities for production of additional food through the practice of
aquaculture. Drainage channels, irrigation canals, reservoirs, holding
ponds, etc., may be features that can be adapted to the production of
fish for food. The inventory of physical resources should consider the
possibility of these features being included. If so, the inventory of
water, soils, and climate should be broadened to consider those factors
important to aquaculture that may be appropriate for the area.

Manmade Facilities

In many proposed project areas, there are existing manmade facilities
such as homes, cemeteries, roads, canals, drains, reservoirs, pipelines
and powerlines, and pumping stations. A complete physical inventory
should be made of any facilities that may be affected by the project or
that are to be incorporated into the project. The project designers
must know the remaining useful life of those facilities to be incorporated and be able to determine the cost and feasibility of improving
them to project standards. People are required to provide the labor
and management needed for operation of the project. There must be
provisions for adequate shelter and healthful living conditions.

Archeological or Historical Resources

There is increasing concern for protection and/or salvage of archeological and historical resources. Much has been willfully or acci-
dentally destroyed in past development activities. There are few new
project activities that will not affect these resources. If they are
assessed early, provisions for salvage, preservation, or abandonment
can usually be made that are acceptable to most concerned interests.
A competent specialist must be engaged to assess the situation in the
area to be disturbed. He should continue to be consulted throughout
planning to assist in identifying practices and features that will
minimize the adverse impact on existing resources of value. It may be
appropriate to abandon some, salvage others, or completely protect
others.

Visual Quality

Visual resources are made up of topography, diversity of geologic
materials, diversity of vegetation texture and density, distribution

25

and visual condition of water in lakes and streams, and compatibility of land use (including farms, forests, and urban areas). An area of high visual quality could include all these items. An area with low visual quality would have predominately one kind of relief, one kind of geologic material, very little water and visually incompatible land uses.

Although drainage and irrigation projects are not normally planned with visual quality in mind, there may be some opportunities that should not be overlooked. A pleasing place for people to live and work is important to the long term success of a project.

Financial Resources

The financial resources of an agency are its tangible and intangible assets which can be used to fund the planning, design, construction, operation, and maintenance of a project in fulfillment of its objectives. Assets may also be categorized under the four "C's" used by financial institutions to evaluate the economic resources of an agency. These are: 1) the character or moral quality of the agency's executive staff; 2) the capacity of the agency's ability to meet payments on time; 3) the capital or net equity of the operations; and 4) the conditions of the money market and the total economy. All are treated in more detail in the following text under sources of revenue, cash reserves, ability and willingness to pay, taxing authority, borrowing capacity, and financial assistance programs.

Projects can be financed by one or a combination of methods. The selection of the best method for financing capital or operational costs requires detailed analysis not within the scope of this manual.

Essentially there are three categories of project costs--long-term, short-term, and recurring. The capital costs of facilities such as dams, reservoirs, canals, or storage tanks are generally large and require financing over a long-term. Short-term loans may be used for smaller items such as pumps, machinery, and on-farm costs. Reserve funds are commonly used for replacement of equipment. Annual revenues such as taxes, fees, charges, and assessments are usually used to defray the costs of operation and maintenance.

Sources of Revenue

Since World War II, the general policy of water project formulation in many countries is to make optimum use of the available resources by incorporating as many purposes into the project as possible, provided each purpose is economically justified (i.e., benefits exceed specific costs). Reimbursable project costs allocated to the respective purposes are recouped in various ways through fees and assessments. Fees are charged for saleable products such as water and power and assessments are made for project services such as drainage, water quality, and other general beneficial purposes. Frequently, fees and assessments are used in combination to provide the total revenue.

Saleable Resources

Saleable resources of a project are the products and services which can be measured and priced quantitatively. Water and power—the two principal products in this category—may be priced in various ways within the legal constraints pertaining to utility industries.

A major regulation which often affects an agency's pricing and taxing policies limits the total revenue to the agency's total operation and maintenance costs plus an allowance representing a reasonable return on the capital investment. This is permitted to encourage the large investments necessary to attain the optimum economic sizing of the project and yet protect consumers against excessive fees and assessments.

Within this constraint, it may be desirable to establish water, power and other service rates with the objective of promoting conservation and making adequate service supply financially feasible for all consumers. Taxes and standby or capacity charges which are alternative sources of supplemental revenue should be initiated as necessary to meet expenses and encourage economic development.

Project Services

Project services result directly from the operation of the project and usually enhance the economic conditions of a geographic area in general in addition to providing specific benefits. Drainage, water quality improvement, and flood control are examples of such services. They normally should be financed with a tax or an assessment to share the costs in proportion to the enhancement of property values. Large public projects which include such services are often financed from general taxes because the services are deemed to be in the public interest.

Cash Reserves

Cash reserve funds are an essential part of any agency's budgeting, if the administrators expect to operate successfully over the long range. Well managed reserves not only enhance the image of the agency's character but also improve both its capacity to meet financial obligations on schedule and its net capital worth.

Reserve funds serve two principal purposes:

o To meet annual payments of indebtedness at fixed intervals.
o To cover emergency and variable O&M costs which are impossible to estimate accurately in the initial budgeting process.

Each reserve account should be funded on a schedule which will culminate with the amount of the desired capital reserve by a designated date. This is to avoid committing the agency's finances prematurely and unnecessarily.

27

Minimum funding for emergency operation and maintenance should be based on some knowledge of and/or historical information about the probability of an emergency occurring or of the variable costs changing. As a rule, the maximum general reserve should be no more than the equivalent of one year's O&M. Funding for an indebtedness or a definite purpose is a matter of scheduling periodic deposits into an account, which will accumulate to the desired amount by a specific date.

All cash reserves except those essential to the agency's operations between accounting periods should be deposited in interest bearing accounts. Public agencies in the U.S. are generally restricted by law to depositing funds into banking or savings institutions where the accounts are insured.

Ability and Willingness to Pay

Where project financing is dependent on sale of a product or service, ability to pay and willingness to pay (or desire) for the project's saleable products and the project services are the basic qualifications to create an effective demand for a project's products and services. Both the ability and willingness to pay must be present simultaneously in a single consumer to be effective because one without the other will not consumate a sale.

Quantitatively, effective demand expresses the amount of a product or services which can be sold at a particular price under the prevailing market conditions. It is the function of many variables such as:

o Price, quality, and dependability of the supply;
o Alternative sources;
o Terms of a contract;
o The future economic conditions;
o Disposable income;
o Agricultural payment capacity.

Planning and feasibility studies used in formulating the project should incorporate an analysis of these variables for the respective purposes and the total project to determine the financial feasibility of each purpose and the project. A purpose and the project are considered financially feasible if the total revenue from the products and/or services exceed the total costs. Further discussions of financial aspects and repayment capacity may be found in Chapter 5.

Taxing Authority

The authority to tax is one of the important prerequisites to obtain public financing for a project. This not only provides more assurance that a loan can be repaid, but also limits financing to federal, state, or lesser political subdivisions. In other countries, project funding may rely largely on the taxing power of the Federal Government and its major political subdivisions.

The more liberal financial programs of federal and state governments in the United States indicate a greater sensitivity to the social well-

being of the population than to the probability of experiencing a
default on the loan. Although taxing authority is still a desirable
prerequisite, it is no longer a restrictive qualification for many loan
programs in the U.S.A..

Borrowing Capacity

Borrowing capacity is an index of an agency's ability to acquire
funding for an irrigation or drainage project. It is predicated on
such basic guidelines as:

- o Assessed valuation in relation to the current debt;
- o Its capacity or ability to meet payments on schedule;
- o The condition of the economy and money supply situation
     in general;
- o Private agencies borrowing may also be affected by the
     moral character of the owner or administrators. Within
     these guidelines, an agency's borrowing capacity is
     interpreted in more specific terms according to the
     lending institutions own policies and criteria.

For example, a private lending institution in the United States may be
reluctant to fund an older established agency's project with general
obligation bonds if the total debt exceeds 15 percent of the assessed
valuation or with revenue bonds if income (in excess of O&M costs) is
not 125 percent or more of the annual payment. Such high risk loans
are more apt to be financed by public institutions such as the Federal
Farmers Home Administration or by a state's resource agency. On the
other hand, a well-managed developing agency with a similar economic
base or financial record may easily acquire funding from a private
financial institution.

Public financing is not ordinarily authorized when private financing is
available at a reasonable cost. As a general policy, all agencies
applying for public loans should be required to submit written proof of
their inability to borrow through private financial institutions. This
will protect the private enterprise and also provide an indication of
the condition of the money market.

Financial Assistance Programs

Financial assistance may be available to public agencies through the
Federal Government for various aspects of planning, constructing, or
operating water projects. Likewise, individual states may have programs
to assist local public agencies.

Federal Government

Various forms of financial assistance for irrigation and drainage
projects may be available through several Federal agencies. The
assistance relates to various project purposes such as water-oriented
recreation, soil and water conservation, community water storage
facilities, development of water supplies, land treatment, flood
prevention, irrigation, drainage, water quality management, and sedimentation control.

29

Because of the numerous and varied federal programs and the possibility of change in programs, this report does not identify specific programs but lists the major federal agencies which have assistance programs. Information may be obtained by contacting State, regional, or local offices of the agencies.

Table 3-1 presents the addresses of the headquarters of the major U.S. federal agencies and the type of assistance the they provide. U.S.A. information may also be obtained from the "Catalog of Federal-Domestic Assistance" which presents a comprehensive listing and description of federal programs and activities providing assistance or benefits. The annual catalog may be obtained through the U.S. Government Printing Office.

State Government

Information on sources of assistance by state governments may be available for the State Legislature or through departments of the State Government involved in financial assistance or agriculture.

International Assistance

At the international level, both multi-lateral and bi-lateral assistance are available from agencies such as those listed in Table 3-2.

TABLE 3-1

U.S. FEDERAL AGENCIES PROVIDING FINANCIAL ASSISTANCE

1. Agricultural Stabilization and Conservation Service
    U.S. Department of Agriculture
    Washington, D.C. 20250

    Programs: Soil, Water, Woodland, and Wildlife Conservation
    Practices

2. Farmers Home Administration
    U.S. Department of Agriculture
    Washington, D.C. 20250

    Programs: Flood Prevention, Water Storage, Land Treatment,
    Irrigation, Drainage, Water Quality Management, Control, Fish and
    Wildlife Development, Public Water-Based Recreation and Water
    Storage

3. Soil Conservation Service
    U.S. Department of Agriculture
    Post Office Box 2890
    Washington, D.C. 20013

    Programs: Flood Prevention, Sedimentation and Erosion Control,
    Public Water-Based Recreation and Fish and Wildlife Development,
    Agricultural Water Management, Rural Community Water Supply, Water
    Quality Management, Pollution Control, Disposal of Solid Wastes
    and Rural Fire Protection

4. Director of Civil Works
    Office of the Chief of Engineers
    Department of the Army
    Washington, D.C. 20314

    Programs: Flood Control Works, Water-Based Recreation, Water
    Supply Storage

5. Division of Trust Facilitation
    Office of Trust Responsibilities
    Bureau of Indian Affairs
    1951 Constitution Avenue, NW
    Washington, D.C. 20245

    Program: Water supply and irrigation facilities to deliver water
    to Indian reservations.

6. Bureau of Reclamation
    U.S. Department of the Interior
    Washington, D.C. 20240

    Programs: (applicable to the 17 western states)
    Irrigation or drainage, or multipurpose projects including muni-
    cipal and industrial water supplies, flood control, fish and
    wildlife, recreation development, and hydroelectric power.

31

TABLE 3-2

INTERNATIONAL AGENCIES PROVIDING ASSISTANCE PRIMARILY
INVOLVED IN PLANNING, RESEARCH, DEMONSTRATIONS,
AND SPECIAL STUDIES

United Nations (Headquarters)
Economic Commission for Europe (ECE)
Economic and Social Commission for Asia and the Pacific
  (ESCAP)
Economic Commission for Latin America (ECLA)
Economic Commission for Africa (ECA)
Economic Commission for Western Asia (ECWA)
United Nations Environment Programme (UNEP)
United Nations Industrial Development Organization (UNIDO)
United Nations Development Programme (UNDP)
United Nations Children's Fund (UNICEF)
United Nations Educational, Scientific and Cultural Organ-
  ization (UNESCO)
World Food Programme (WFP)
World Health Organization (WHO)
World Meteorological Organization (WMO)
International Labor Office (ILO)
International Atomic Energy Agency (IAEA)
Food and Agriculture Organization of the United Nations (FAO)
Organization of American States (OAS)
International Bank for Reconstruction and Development
  (World Bank) (or, IBRD)
Inter-American Development Bank (IDB)
Asian Development Bank
African Development Bank
Arab Fund for Economic and Social Development
Islamic Development Bank
Kuwait Fund
Organization for Economic Cooperation and Development (OECD)
European Investment Bank (EIB)
Also, many Bi-lateral Programs with various countries.

Human and Institutional Resources

At all stages of irrigation and drainage project development consideration should be given to the human and institutional resources. In
earlier stages such resources must be mobilized to insure that the
project which is formulated is well conceived and possesses a high
likelihood of success. In later stages such resources must be mobilized to construct, operate, and maintain the project works and ancil-
lary developments within the community and the region.

As with the physical and financial resources, the human and institutional resources are inventoried, analyzed, and synthesized during
project formulation. Relevant considerations include geographical
location, existing qualifications, and receptivity to change and
development. Somewhat different emphases are later placed on these
activities depending on the stage of project development, i.e. precon-
struction, construction, and post-construction (or operational) periods.
Inadequate institutional resources during any of these periods will
jeopardize the success of the irrigation and drainage project.

In order for an irrigation and drainage project to achieve its objectives, human and institutional resources must be well used. During the
project formulation process consideration must be given to the quantity
and quality of such resources which are required for successful operation of the project. Human resource considerations include: 1) the
number of people whose lives will be significantly influenced by the
project and 2) their qualifications to farm the lands of the project
and to operate and maintain project facilities. Institutional considerations include: 1) the organizations which are required to provide
inputs to the agricultural community which is served by the project and
process, store, and transport project products and 2) the laws,
regulations, and customs which form the setting and provide constraints
on how the project objectives are achieved.

Human Resources

Irrigation and drainage projects are formulated to serve the needs and
wants of people. Such projects generate employment and provide goods
and services. The result is economic development and improved quality
of life.

Number of People Served

Beneficiaries of projects can be classified into five groups:

o   Those who farm and own the land;
o   Those who operate and maintain project facilities;
o   Those suppliers whose employment stems from providing inputs
    to the farming operations and, to a lesser degree, to the
    operation and maintenance of project facilities;
o   Those who process, store, and transport project products; and
o   Consumers of project products.

Specific project objectives will usually include some combination of
the following:

33

o   Improve economic development of the region in which project
    lands are located,
o   Improve standard of living for the farmers,
o   Maximize the number of project farmers,
o   Improve variety and quality of food and fibre for consumers,
    and
o   Favorably influence the country's balance of payments either
    by increasing exports or by import substitution.

Early in project formulation an appropriate set of specific project
objectives should be selected. The selection process should include
interrogation of the project decision makers, although definitive
statements may not ensue. The potentially significant impact of project
objectives on project formulation may not be recognized at the initial
planning stage. Therefore, the planner should be prepared to review
and modify his initial assumptions on project objectives at later
stages of planning.

The introduction of project objectives here is to call attention to the
relationship between objectives and the number and incidence of people
who are served by the project. Early in the project formulation process
decisions must be made regarding farm organization. Such decisions are
a logical farm size, crops to be produced, and the social organization
of the farmers.

Simultaneous consideration of project objectives and identification of
adapted crops will lead to an appropriate cropping pattern. Agronomic
and economic analyses of the cropping pattern will establish the likely
range of net income per unit of land area. Farmer income objectives
will then be used to establish land area per farmer. A knowledge of
the land area available for project development will then establish the
number of farmers to be served. The number of farms in the project is
a function of the social context and will be discussed later under
institutional resources.

The number of people required to operate and maintain project facilities
is determined by analyzing the requirements of the particular facilities. For most requirements, alternative methods and equipment are
available. Selection of the level of automation will establish the
trade-offs among number of O&M personnel, dependability of operation
under project conditions, and capital investments. Guides for establishing the number of people involved are found in the ASCE Manual of
Engineering Practice, "Operation and Maintenance of Irrigation and
Drainage Projects."

In most cases the formulator of irrigation and drainage projects is not
required to set precise values on the number of people who will be
served by the project by providing inputs; who will process, store, and
transport project products; and who will be consumers of project
products. Where such estimates are required it can prove helpful to
study the development of similar projects in the same or similar
regions.

Qualifications of People Served

Project formulation includes the identification of the people who will farm the land, the level of their qualifications as a result of previous practice and training, and the formulation of training needs. Training needs include not only the identification of the particular skills and knowledge which must be transferred but also the institutions which will provide the training. Usually consideration must be given to several different levels of training, such as extension workers, vocational/technical institutes, and colleges. In the United States the resources of agricultural extension staffs have been very effective in training farmers. In some developing countries similar programs are feasible. "The Training and Visit System" is an approach to agricultural extension education which has proved to be beneficial in developing countries.

Efficiency of Resources Use

In planning an irrigation and drainage project it is customary to consider the efficiency with which the water resources will be used. Alternative methods of water distribution and application result in varying efficiencies. The efficiency concept is also relevant to planning for the use of the human and institutional resources. For each alternative formulation of the project one can conceptually relate the number of people and qualifications thereof which are required by the project to the number of people and qualifications thereof which are committed to the project. Defined in this manner, a high ratio would suggest an efficient use of human resources. If a specific objective of the project is to assist in providing employment, a low ratio would be desired. An analogous concept is useful when evaluating the appropriateness of existing or proposed institutional resources.

Institutional Resources

The objective of the institutional analysis is to 1) inventory the existing institutions, 2) determine their applicability to the project being formulated, and 3) propose new institutions or modifications to existing institutions as may be needed to enable the objectives of the project to be achieved. Where institutions need to be established or modified it is important that they conform with the needs, cultures, and desires of the people that they will serve.

If an irrigation and drainage project is to succeed extensive changes must occur in the community. If these changes do not occur rapidly project beneficiaries are either not being served or are being served to a lesser degree than was projected. Furthermore, with high discount rates rapid movement to with-project levels of production is necessary for the project to give the projected, economic return. Experience suggests that for rapid and extensive change to occur that some kind of widespread enthusiasm or excitement is apparently needed. Traditionally, in the United States the opportunity for self-advancement has been the driving force. Other generators of enthusiasm which may be relevant include religious, revolutionary and nationalist movements.

The services of several kinds of institutions are required for the successful development of an irrigation and drainage project. They include:

o   Planning, design, and construction organizations encompassing the disciplines of agriculture, engineering, economics, social science, and environmental sciences;
o   Applied research and extension education organizations;
ɔ   Suppliers of production inputs such as seeds, fertilizers, pesticides, machinery, fuel, and lubricants;
o   Suppliers of capital for various loan durations such as long term for real estate, intermediate for machinery, and short term for operating costs;
ɔ   Marketing services such as grading standards and inspection, accessible markets, market information, and transport;
ɔ   Processing and storage of agricultural products; and
ɔ   Laws and regulations governing farm size, transfer of land, and allocation of water.

During project formulation institutions will be identified which will implement these functions. Where institutions already exist the analysis will determine their adequacy to meet the needs of the new project. In some cases existing institutions will need to be modified; in others new institutions will need to be formed.

Water Allocation

Because water supplies are finite and because there are competing uses for the limited supplies, project formulation will include consideration of how the available water is to be allocated. The objective of water allocation is to resolve conflict among water users both within the irrigation and drainage project and between the project users and other users.

Farmers view allocation rules as constraints on the availability of water. They adopt water use technologies which are economically consistent with the institutional constraints. Project formulators are obliged to review any prevailing water allocation rules to determine their impact on project feasibility. After undertaking such a review they may find it desirable to propose changes in the rules which will either increase the likelihood of achieving a certain level of project benefits or permit using a higher level of benefits in the project analysis.

Number of Farms

The social organization of a farming community influences how farm work is accomplished. A given project land area could consist of many operating units of small size or a few operating units of large size. Each operating unit could use the services of one or several farm workers. Answers to both of these questions need to be developed during project formulation by considering the economics of production and the specific project objectives. Once the farm size is selected, the number of farms is a function of the project service area.

36

Farm Size and Transfer

Farm size in an irrigation and drainage project is determined by factors such as:

o   Historical ownership and operating unit patterns in and near the land of the new project,
o   The economics of production of adopted crop and livestock operations,
o   Labor intensity and level of mechanization of the projected cropping pattern, and,
o   Social/political objectives of the decision makers.

Other institutional constraints may also influence farm size. Consider, for example, the impact of the U.S. System of surveying the Public Lands on the sizing of irrigation farming units. The dominant farm sizes in the United States are 160 acres or some fraction or multiple thereof.

Historical patterns tend to dominate when the new project encompasses lands which previously have been used for dryland farming. When the project is to bring new lands into production there is an opportunity for farm size to be selected to satisfy social/political objectives.

There is considerable merit to adopting a flexible stance on farm size. The uses of the lands will probably change with time. The new uses may be either more intensive or extensive than the old uses suggesting either a decrease or an increase in farm size. New machines may be developed which lead to greater net returns per unit of area with larger farms. A change to more intensive land use and uses requiring greater labor inputs may make smaller farms more economically viable. A flexible stance implies the ready transfer of land ownership or the rights to farm the land. It may also mean some institutional constraints on the disposition and transfer of farms at times of death or retirement of the farmer in order that the continuation of economically sized farm units may be assured.

References

1.   "A Framework for Land Evaluation, Soils Bulletin 32," United Nations, FAO, Rome, 1976, (M-51 ISBN92-5-100111-1).

2.   "Economics and Soil Science-Copartners in Land Classification," D. Nielson, Agricultural Economist. Paper presented at Bureau of Reclamation Land Classification Meeting in Region 7, February 12, 1963.

3.   "Irrigation, Drainage and Salinity, An International Source Book," United Nations, FAO/UNESCO, Hutchinson & Co. LTD (Publishers).

4.   "Long-term Planning of Water Management, Volume II," United Nations, 1976.

5. "Soil Conservation Service Guide for Environmental Assessment," U S. Department of Agriculture, March 1977.

6. "The Training and Visit System," Benor, Daniel and Harrison, James Q., Agricultural Extension. World Bank. May, 1977.

CHAPTER 4. FORMULATION OF ALTERNATIVE PLANS

Introduction

Need for Alternative Plans

The fourth step of the formulation process is to develop alternative plans to meet the identified problems and needs through developing and managing the available resources. The formulation of a single plan may be sufficient when there are few or no constraints and where objectives are complementary or noncompetitive (i.e., the satisfaction of one objective contributes toward or does not preclude the satisfaction of the other objectives). However, in most cases, alternative plans have to be developed. There will usually be competition or conflict between objectives as the achievement of one may reduce the achievement of the others. Further, a single specified level of satisfaction of need for a given objective may not be acceptable through time due to uncertainty of future conditions. Other factors which contribute to the need for formulating alternative plans include limited resources, technical planning constraints, economic and financial constraints, acceptability, and legal, institutional, and administrative constraints.

Iterative Process

Plan formulation is an iterative process. A large number of possible plans should be initially considered, and at this point the level of detail is necessarily general. At each iteration, the number of plans considered usually will be reduced and the level of detail increased until finally one or two or three plans will remain for final analysis. The level of detail of analyses of the final list of plans must be such that any plan could be selected for implementation. It is expected that only a few plans will be analyzed in detail.

Planning Situation

Plans must be tailored to the planning situation in the local area. The physical, financial, institutional, and human resources described in Chapter 3 will vary considerably from one area to another. Plans which are highly viable for one setting may be completely inappropriate for another location. Projections of future conditions must be considered as well as existing conditions.

Evaluation Parameters

Plans selected for detailed evaluation should potentially meet the tests for effectiveness, completeness, efficiency, and acceptability as described in Chapter 5. Although it will not be known until after detailed analysis whether a plan in fact meets all four tests, plans known to fail one or more tests should not be carried to detailed analysis.

Formulation of Plans

Public Participation in Plan Formulation

In the plan formulation phase of developing irrigation and drainage projects, interaction or dialogue types of activities should be used if a public participation program is necessary. An active exchange of viewpoints about constraints, priority concerns, and possible implications will broaden the range of alternatives to be considered. It will also develop a better understanding of their technical, economic, social, and political feasibility.

Role of Creativity and Innovation

Creativity and innovation should play a major role in plan formulation particularly in view of rising construction costs and, in some developed nations opposition among some factions of the public to new large, structural projects. By taking a water management approach at the onset of a planning investigation, more comprehensive, more efficient, lower cost, and more acceptable alternative project plans can be formulated. A water management approach includes the systematic study and evaluation of existing and potential use within a study area to determine how present and near-term water, drainage, and related land resources needs can be best met.

A water management approach should be taken at the onset of all planning studies regardless of the types (structural, nonstructural, etc.) of alternative plans expected to be formulated.

In addition to traditional surface water developments, other water sources which should be considered include ground water, municipal waste water, and agricultural return water. Desalination of brackish water or seawater and importation of water from other basins are other potential supplies. Vegetation conversion by removal of brush and planting grass may increase water yield by reducing evapotranspiration. Weather modification by cloud seeding may also increase water yield under special circumstances. Improved management techniques include the conjunctive use of several water supplies, municipal water conservation, industrial water reuse, and agricultural water conservation through changing cropping patterns, recycling or reuse of agricultural return flow, improved on-farm systems and scientific scheduling of irrigation. Water exchanges and dry-year deficiencies can extend the yield of existing and potential projects. Low cost alternatives, such as temporary diversion dams designed to withstand only low frequency floods, should also be considered as alternatives to expensive permanent facilities. Labor intensive alternatives, even if not the most cost effective, should also be considered if regional employment or improvement in social conditions are important project objectives.

Post Analysis

Comparing and evaluating the performance of existing irrigation and
drainage projects with the objectives that were envisioned for them can
provide valuable information to the plan formulator. This is known as
post analysis and involves comparison of the actual or post project
costs, benefits, and other projections with the original estimates.
With the exception of required evaluations conducted by the World Bank,
post analysis seems to have received little attention in public works
projects undertaken by governmental agencies and almost no attention by
private firms.

In conducting post analysis, one must be careful to evaluate the project
in terms of societal goals and objectives that existed when the project
was planned, and not in terms of those that have occurred since.
Defining these goals and objectives is the first step in conducting
post analysis.

The second step is data collection, including data on what was projected
and what actually happened. Physical data include hydrologic, geologic,
soils, topography, environmental data, and climate. Financial data
include cost estimates, cost sharing programs, repayment plans, and
financial arrangements. Economic data include estimated benefits from
the project. Institutional data include information on the institutions
set up to implement and manage the project and how well they have
functioned. Legal data include laws in effect and which changed during
the planning, construction, and operation of the project and any legislation that was required to implement the project. Technical data
include design data for project facilities and operation and maintenance
plans. Other data include projections of population, water use, and
area development.

The third and most difficult step in post analysis is data evaluation
to determine the magnitude of the disparity between forecast and actual
data, and to determine the cause of the disparity. Projected and
actual conditions must be compared and evaluated in terms of the goals
and objectives that were established when the project was planned.

The main lesson to be learned from post analysis is that the future
will most surely be different from the projections that are made. The
goal for the planner, therefore, should not only be to minimize the
difference between the projected and the actual but also to plan for
it. By conducting post analyses on existing irrigation and drainage
projects the planner can learn where and how previous planners had gone
wrong, and can hopefully avoid making the same mistakes.

Define Project Elements

Once the entire list of alternative project plans has been developed,
the features of each alternative must be defined so that a screening of
plans can be made prior to detailed plan analysis. The location, size,
and type of structural measures must be defined so that rough cost
estimates may be made. Cost curves for both construction and operation,
maintenance, and replacement costs can be used since incremental costs

41

among plans are more important than total costs of alternatives at this point. Construction procedures and schedules should be predicted, as well as requirements (techniques, equipment, and personnel) for project operation, maintenance, and replacement. Similar data must also be developed for nonstructural components of alternatives. The legal, institutional, and social aspects and constraints of all the alternative plans must also be defined. All features of the alternatives should then be arrayed or organized for easy comparison among the alternatives.

Benefits and Costs

Economic and Financial Feasibility are covered in detail in Chapter 5, Plan Evaluation. In the plan formulation stage, a rough estimate of project benefits should be made for each plan to be compared with project costs for that plan. Methods of comparing costs and benefits, also described in Chapter 5, include the benefit cost ratio, the internal rate of return, and the net present value methods. The benefit cost evaluation is an important part of the process for selection of alternatives to be evaluated in detail.

Phasing, Staging, and Timing

Phasing, staging, and timing of each alternative plan must be carefully considered to coincide with meeting the needs and objectives of each plan while avoiding large investments of resources before they are needed, inducement of too rapid growth, and the foreclosure of options for the future. The time when revenues and other project benefits will occur must be estimated to determine their effect on project economics. For example, a benefit or return which does not occur for 20 years must be greater than $4 million to justify a current capital outlay of $1 million if the discount rate is 7 3/8 percent.

Providing a water supply too soon may also induce rapid, unwanted growth or encourage excessive application of water. For example, development of an irrigation water supply long before it is needed may induce urban growth into the agricultural area which would defeat the project objective. On the other hand, projects may need to be implemented as soon as possible to achieve economic growth, employment stability, and social goals.

To avoid foreclosing options for the future, an alternative future planning concept may be used. This concept seeks to protect and create options for the future by identifying high cost decisions which may preclude a desirable future. An example of a high cost decision is a decision to rely on reclaimed waste water as the sole source for making up water deficiencies. While waste water may be adequate for agricultural use, it could "kill off" a future in which the water was needed more for municipal use rather than agricultural. Sometimes apparently modest decisions turn out, upon analysis, to be very high cost. Other times, alternative plans can be designed to meet needs while minimizing or avoiding high-cost decisions.

42

It is important in the phasing and timing process for the alternative plans to identify those decisions which are high cost and if they are unavoidable, to stage them as far in the future as possible.

All of the above reasons for staging and delaying alternative plan features must be carefully weighed against reasons for early plan implementation, including favorable political climate, public acceptance, availability of funding and other resources, and the rate of inflation. Phasing, staging, and timing may in itself create additional alternative plans.

Identify Constraints

Identification of constraints should be done early in the plan formulation process to avoid pursuing projects which are not viable because of physical, economical, financial, institutional, or human constraints. Many of these constraints can be identified by initially reviewing the available resources, described earlier in this report, which must be inventoried prior to formulation of plans. For example, review of the physical resources may identify limitations of the soil for crop growth, water quality limitations for growing certain crops, and climate limitations such as a very short growing season.

Environmental constraints must also be considered, particularly those mandated by national, state, or local laws. Measures to mitigate adverse impacts on fish, wildlife, and other environmental resources may be required and must be considered as a constraint. The cost of such mitigation measures must be included in the economic evaluation of project plans.

Of the human constraints, need for relocation and resettlement of persons is of critical importance. Resettlement costs of $10,000 per family are not unusual, even in very poor rural areas. Such costs should be included in project costs and provided for in financing arrangements. Involuntary resettlement generally is, and certainly always potentially is, a politically sensitive measure, giving rise to special social and technical problems. It is important that a resettlement plan be developed in considerable detail during the feasibility study phase. Such a plan could include some level of compensation as one element, but also should include definite locations and alternative locations, and facilities to help ensure that settlers are offered opportunities to become established and economically self-sustaining in the shortest possible period, at living standards that at least match those before resettlement.

Time can also be an important constraint. The time required to plan, design, authorize, fund, and construct a complex, structurally-intensive project may not meet critical short-term needs. A less complex, more readily implementable project may be called for to meet short-term needs. Timing must also be considered as it relates to other constraints such as a short construction season or a low flow period for constructing diversion works.

Once the constraints are identified, they should be arranged or displayed in an organized manner so they may be readily applied to the various alternative plans. As plans are formulated and checked against the constraints, additional constraints may be identified and should be added to the list. Plan modifications to fit within the constraints will also become obvious, thus creating additional, more viable plans.

Develop Initial Alternatives

The initial list of alternatives should be developed without screening or ranking based on cost or other constraints. It is important at this point to develop a complete list, because many nonviable proposals may have significant public interest and support. It should be well documented that these plans were considered and justification given for not selecting them for further analysis.

Many subalternatives of plan components may be identified if the planning area is large or if more than one objective is being considered. Several components would then be combined to form an alternative plan which would serve the entire planning area. If several components are identified, they could be combined in different ways to form a very large number of alternative plans to meet different planning objectives or various combinations of objectives. At this point, plans should not be developed in detail but should be general in scope.

The public participation program can be used very effectively in developing plan components and entire alternative plans. A brainstorming session can be held to generate an initial "idea list." This list can be refined into plan components for presentation and discussion at a subsequent session. New ideas and revisions to the components can be developed at a second session. The components can then be combined to form several initial plans to serve the entire study area. One plan should be formulated which emphasizes each of the objectives, and one or more plans should be formulated to meet a mixture or combination of objectives. The plans can then be reviewed, discussed, and revised at a third public participation session. The result will be a large number of alternative plans with different objectives reflecting the ideas and values of different sectors of the public. Care is required to prevent local interests from playing one agency or institution against another to get the best cost sharing rather than the most economic and environmentally optimized plan.

Formulation of Operation and Maintenance Plans

In the plan formulation stage, operation and maintenance (O&M) plans should be formulated in concept to determine if any of the plans have extraordinary O&M problems or costs. The type of governing body of the organization which will operate and maintain the project needs to be considered, as well as the necessary staff. If specialized skills or training are required, their availability must be determined. High operating costs, such as energy requirements, must also be considered. Special problems such as safety and water quality may also require expensive and difficult operations and maintenance procedures. Legal requirements may also affect operations.

The best guidance on O&M plans can be obtained by evaluation of O&M experience at existing similar projects.

44

Selection of Alternatives to be Evaluated

The entire list of potential alternative project plans must now be reduced to a workable number of plans for detailed plan evaluation, selection, and optimization. A screening process should be used to identify the plan or plans which best meet the identified problems and needs. A future without a project plan must also be described for comparison with the selected alternatives.

Each alternative project plan most be compared with the lists of constraints, identified problems and needs, alternative plan accomplishments and elements, and phasing and staging tables. Redundant and obviously inefficient and unacceptable alternative plans can be readily eliminated. Each alternative plan selected for evaluation should meet the tests for effectiveness, completeness, efficiency, and acceptability defined in Chapter 1 and described as part of Step 5.

Beneficial and adverse effects must be treated comparably when relating water and land resource plans to one another. All alternative plans displayed should meet these four tests in the event a plan other than the plan initially recommended is selected for implementation. Where difficult tradeoffs are required between alternative plans, it will be difficult to determine when these criteria have been adequately met.

In describing future conditions without a plan, developments expected to occur in the area of study without an irrigation and drainage project should be identified and considered under operation. These developments would include municipal and industrial water, power, agricultural irrigation, drainage, recreation, and fish and wildlife enhancement, and management programs to obtain flood control, pollution abatement, and other needs. Environmental and social enhancement or degradation that will occur to the area of study should be recognized. Care must be used that the without-plan does not include development based on anticipation of the project. This may reduce economic benefits of the project, but may also reduce the mitigation requirements. Although future conditions without a plan are important in determining the net effects of the various alternative plans, the future without plan does not always need to be considered or displayed as an alternative plan itself. Special conditions may prevail in which public controversy requires nondevelopment to be analyzed even though it is not a "plan" and does not meet the tests of viability with regard to projected needs such as irrigation water supply. This may be required when nondevelopment proposals must be carefully analyzed in order to demonstrate the relative value or utility of development-type programs and projects.

Considerable judgement must be exercised in selecting the future without project condition. To decide whether or not a development should be included in the future without project condition, there should be ample evidence that these developments will be implemented. Strong community support and economic and financial advantages are just some of the criteria that should be considered in deciding if the potential action should be included in the future without project condition.

The selection of alternatives to be evaluated is a reiterative process, and several iterations through the above steps will usually be required to fully consider all alternative plans and to make the final selection of alternatives for detailed evaluation, comparison, and optimization.

The planner should make an effort to limit the list of alternatives to those that are truly alternatives. Expanding the list with alternatives which have only subtle differences not only increases the cost of analysis but also adds unnecessary complexities which make understanding and actions by decision-makers more difficult.

Optimization and Energy Considerations

Optimization

A primary goal of the plan formulation process is to optimize, from an engineering, economic, environmental, and social standpoint, the use of available water and related land resources to fulfill identified objectives and needs. From an engineering and economic standpoint, production theory, operations research techniques and sensitivity analysis can be utilized to accomplish this goal. However, because a project's environmental and social aspects may be primarily nonmonetary in nature and involve considerable subjective analysis, no practical economic or mathematical models have been developed for the planner to utilize in comparing the noncommensurate economic, environmental, and social aspects of a project. The major concept or techniques utilized to reach consensus or gain this acceptance is the acceptability test which is discussed in another section of this report.

Marginal Analysis

Concepts associated with production theory can be utilized to optimize a project's engineering and economic aspects. Production theory is concerned with the relationship between project inputs and project outputs and utilizes concepts related to marginal analysis.

Marginal analysis is based on the theory that optimum development is obtained when the last dollar's worth of input or cost results in just a dollar's worth of output or benefit. If a marginal dollar of cost results in more than a dollar of gross return, the net return can be increased by increasing the investment. The maximum or optimum net return will be attained when the marginal cost equals the marginal benefit. The following example demonstrates the concepts discussed above. In this example, the most significant relationship to consider is the relationship between total project benefits and total project costs, as illustrated in Figure 4-1. Four possible scales of development are represented by Points 1, 2, 3, and 4. Point 1 is the scale of development at which the ratio of benefits to costs increases from less than unity to greater than unity. Point 2 represents the scale of development at which the ratio of benefits to cost is the greatest. Point 3 is the scale at which the benefits exceed costs by the maximum amount. Point 4 is the scale at which the project benefits again equal projects costs.

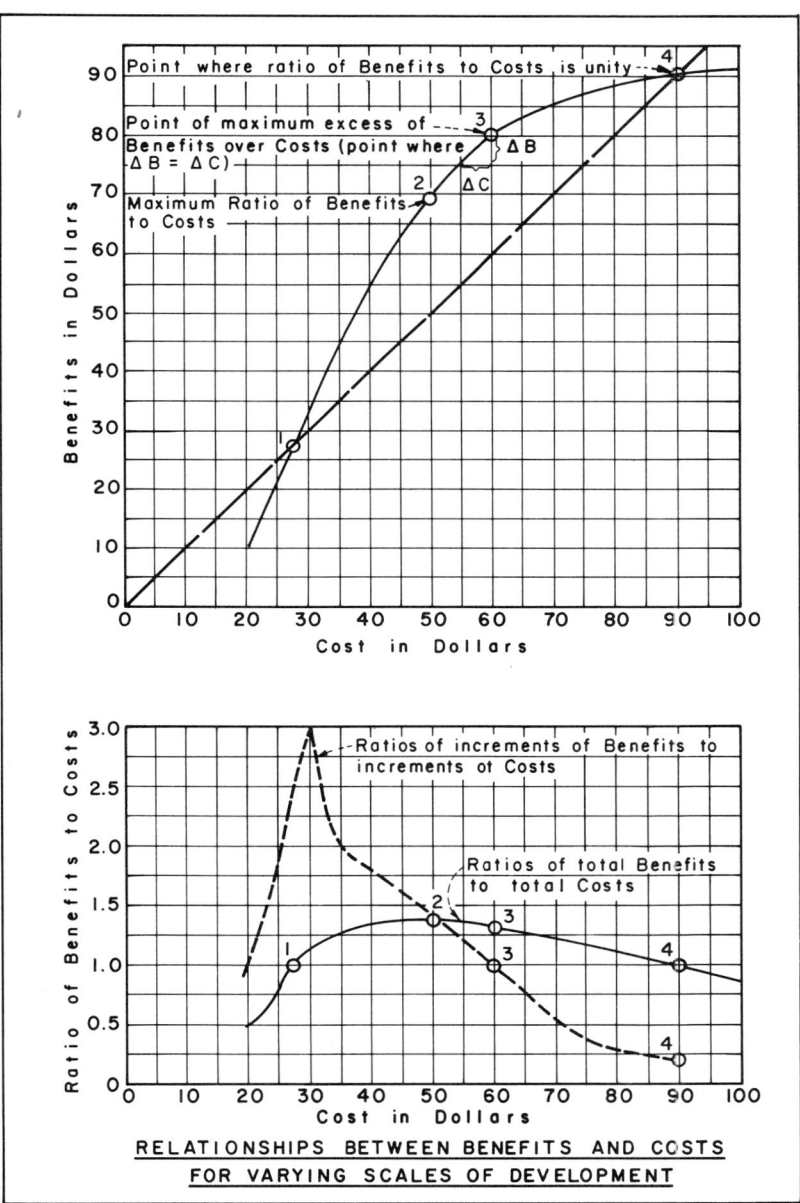

RELATIONSHIPS BETWEEN BENEFITS AND COSTS
FOR VARYING SCALES OF DEVELOPMENT

Figure 4-1

If the scale of project development were established at Point 2, the rate of benefit accrual per unit of cost would be at a maximum, but the full economic possibilities of the site would not be utilized. There would remain additional increments of development for which the benefits would exceed the costs. These additional justifiable increments lie in the zone between Point 2 and Point 3.

At Point 3, the cost of adding the last increment in scale of development is equal to the added benefits resulting from that increment. At this point, we have the maximum excess of benefits over costs.

Between Point 3 and Point 4, although the overall ratio of benefits to costs is unity or greater, the benefits added by each increment in scale of development are less than the cost of adding that increment. Extension of the scale of development into the zone beyond Point 3 is not economically justified.

Incremental Analysis

Marginal analysis implies that continuous increments susceptible of infinite division will be analyzed. Since irrigation and drainage projects, except in special cases, cannot be represented by a continuous function, this approach is not practicable, and therefore, a modification of the marginal analysis approach, generally referred to as incremental analysis, is utilized in the plan formulation process. Incremental analysis deals with larger discontinuous increments of size, scope, or means of development on which there is a practical choice for inclusion in, or omission from, the plan.

The segments or increments of scale of development to be considered are the smallest on which there is a practical choice for inclusion in or omission from the plan. An increment can be one of a group of projects in a basin or subbasin plan; or it can be a part of an individual project such as a purpose added to a single- or multiple-purpose use of water or of a facility; a difference in scale or intensity of water use for single or multiple purposes; a difference in degree of water shortage allowed for in the development plan; or a difference in size of one or a group of facilities, etc. Typical increments of analysis may also include differences of reservoir capacity, irrigable area, installed power-generating capacity, amount of water developed, drainage collector systems, etc.

In evaluating an increment of scale of development, the differences in the benefits and in the costs "with" and "without" the increment included in the plan are the benefits and costs attributable to that increment.

The principles and criteria described above apply to plans formulated for regions, river basins, subbasins, individual projects, or divisions of projects. To be justified for inclusion in a plan, each project in a group and each segment or increment of a project should add more benefits than it adds costs, and there should be no more economical means of development (water resources or otherwise) that would be precluded from development if the project or segment were undertaken for accomplishing equivalent benefits and purposes in the area.

Thus, the engineering and economic aspects of plan formulaticn investigations and analyses resolve largely down to benefit and cost evaluations
of alternative plans and comparisons of differences in the benefits
and costs. The differences in alternative plans in this connection
would be in scale and intensity of development and in physical means
of development. The principles and concepts involved in methods of
incremental analyses are particularly useful tools in determining the
most economical means and scales of development of the various alternative plans.

Operations Research Techniques

Although marginal or incremental analysis is the primary technique
utilized in formulating irrigation and drainage projects, considerable
attention has been directed in recent years toward the application of
operations research techniques to the field of water resource planning.

Linear programming, dynamic programming, simulation analysis, non-
linear programming, integer programming, queuing theory, netwзrk
theory, gradient search techniques, and combinations of the aэove
optimization techniques, have been applied to water resource development
programs.

Although a discussion of the above techniques is beyond the scope of
this report, the basic concepts behind the use of these techniques are
the same.

Initially the planner determines if the water resource problems can be
represented by a model. If the answer is affirmative, applicable
models are analyzed and one, or in some situations more than one,
model is selected. Next the dependent and independent variables,
constraints and the equations needed to define the system are selected
and developed. Finally the basic data and operating criteria needed
to quantitatively define the system are developed. Limitations on the
availability of data frequently limit the effectiveness of models.

Simulation analysis, linear programming, and dynamic programming, and
combinations of the above techniques are the most common operation
research techniques used today in water resource planning.

Sensitivity Analysis

An important technique used in conjunction with incremental analysis
and operations research techniques is sensitivity analysis. Through
sensitivity analysis, the planner tests key parameters to determine
how sensitive an alternative's outputs are to changes in the selected
parameters. Sensitivity analysis is especially useful where conflicting
viewpoints or considerable uncertainty are involved, such as with
water supply and in projecting future needs. Both optimistic and
pessimistic values, within a realistic range, should be used to test
key parameters.

Sensitivity tests are used to identify which costs and benefits have
significant effects on the project economic or financial viability, as

their values are varied, and the extent of these effects. They help in
identifying design or analytical factors on which, perhaps, more
attention should be focused. They are likely to be most effective
during project formulation. Sensitivity tests do not, by themselves,
take account of uncertainty, but normally should precede risk analysis.
Table 4-1 is a common form of summary presentation of sensitivity
testing. Again, electronic computers aid greatly in such analyses.

Table 4-1

Sensitivity Tests on Economic Rate of Return

		Economic	Rate	of Return	
			Base		
Economic Product Prices	-20%	-10%	Case	+10%	+20%
Costs:					
Base Case	9.7	13.5	17.0	20.0	22.9
Capital Costs down 10%	11.2	15.3	18.9	22.1	25.1
Capital Costs up 10%	8.3	12.0	15.3	18.3	21.0
Operating Costs down 10%	11.4	15.1	18.4	21.3	24.1
Operating Costs up 10%	7.8	11.9	15.5	18.8	21.7
Capacity Utilization down 10%	7.5	11.3	14.6	17.6	20.4
Delay of Completion (1/2 year)	9.1	12.6	15.8	18.6	21.1
Base Case without Shadow					
Exchange Rate	5.0	8.7	12.2	15.3	18.1
Growth Rate of Output					
Demand in the Northeast down 50%	9.2	13.0	16.5	19.6	22.4

Recently, more economists are recommending the use of "switching
values," rather than the preceding presentation. The switching value
of a variable is that value at which the project would still remain
viable. For example, the switching value of the yield per hectare
(ha, 10,000 square meters or 2.471 acres) of an important crop in an
agricultural development would be its lowest value, other things
remaining equal, at which the project would remain acceptable. In
Table 4-2 the switching value is given as the percentage change which
would reduce the Net Present Value (NPV) to zero at the Opportunity
Cost of Capital (OCC) (or the Internal Rate of Return (IRR) to the
OCC)

Table 4-2

Sensitivity Analysis

Variable	As Appraised	Switching Value	Percentage Change
Yield (metric tons/ha)	4	3	-25
Construction Costs ($/ha)	4,250	5,950	+40
Irrigated Area per Pump (ha)	50	25	-50
Standard Conversion Factor	0.85	1.36	+60

50

In this example, the yield is clearly the most critical variable, as the switching value involves the least percentage change. If experience suggests that the yield can easily be that much less than expected, perhaps because of lack of effective extension services, then this project would appear to be quite risky unless the project were to include a strong extension improvement program. The project is also sensitive to construction costs, but a 40 percent increase in these costs (in real terms) may be considered quite unlikely if the state of engineering design and studies for the project is well advanced. Correlation among the variables can pose serious problems. If variables are highly correlated, the related variables must be varied jointly.

Risk Analysis

While risks normally have been mentioned in specific sections of feasibility reports for irrigation and drainage projects for some time, and specific risk factors have been frequently used for spillway design and flood control benefits, it is only lately that some agencies are urging that specific risk analyses be made on an overall project basis.

Situations of risk are defined as those in which the potential outcomes can be described in reasonably well-known probability distributions such as the probability of particular flood events. Situations of uncertainty are defined as those in which potential outcomes cannot be described in objectively known probability distributions.

Risk and uncertainty arise from measurement errors and from the underlying variability of complex natural, social, and economic situations. Methods of dealing with risk and uncertainty include:

   o   Collecting more detailed data to reduce measurement error
   o   Using more refined analytic techniques
   o   Increasing safety factors in design
   o   Selecting measures with better known performance characteristics
   o   Reducing the irreversible or irretrievable commitments of
       resources
   o   Performing a sensitivity analysis of the estimated benefits
       and costs of alternative plans

Reducing risk and uncertainty may involve increased costs or loss of benefits. The advantages and costs of reducing risk and uncertainty should be considered in the planning process.

The planner's primary role in dealing with risk and uncertainty is to identify the areas of sensitivity and describe them clearly so that decisions can be made with knowledge of the degree of reliability of available information.

Energy Considerations

Although the cost of energy for project use is taken into account in the plan formulation process and in design of project facilities and is explicitly accounted for as an operating cost, rapidly escalating energy costs require those involved in the planning of water resource projects to become more cognizant of its value.

Currently, power producing and consuming facilities are formulated utilizing energy costs and benefits based on current free-market values. However, in light of each nation's emphasis on energy conservation and the rapidly escalating cost of energy, the planner should, on a case-by-case basis, consider other plan formulation concepts when energy using or producing facilities are involved. Adjusting future energy costs for expected inflation, or developing energy budgets, are concepts that should be considered.

With respect to energy budgets, there has been much discussion about analyzing a project's energy production and consumption through the development of an energy budget. Acceptable methodology is currently being developed. As a minimum, the planner should determine each alternative's sensitivity to energy costs.

CHAPTER 5. ECONOMIC AND FINANCIAL ANALYSES

Introduction

Complete economic and financial analyses of irrigation and drainage
projects are complex processes needing the inputs and guidance of
persons especially experienced and trained in these important aspects
of plan evaluation. Normally such persons are professional economists
and financial analysts with specialized experience with these types of
projects. But these analyses cannot be prepared by such persons alone.
They need the inputs and advice from others who are experts in the
various other technical aspects involved in the projects. For irrigation and drainage projects these usually would include engineers,
agronomists, land classifiers, soil scientists, and other specialists,
as appropriate.

This chapter provides a summary of the more important principles of
economic analyses, and to a much less extent of financial analyses, for
irrigation and drainage projects. It is directed towards the non-
economist. A limited number of hypothetical examples are used to
illustrate some points. Examples of complete project analyses are not
included because of space limitations. However, several examples are
contained in the references listed at the end of this chapter. Furthermore, many agencies have specified format and analytical requirements
for projects under their cognizance. Such guidelines should be obtained
early in the planning process for projects which those agencies may be
called upon to review. It is believed that if one understands the
principles presented herein, he should be able to follow any agency
guidelines for project economic and financial evaluation with understanding and without undue difficulties.

Economic Feasibility

Irrigation Projects and the Economy

Judgments as to the economic feasibility of a project should be based
not only on a narrow economic criteria for the project itself (micro-
economics), but also on some larger considerations as to the role of
the project in the overall development of a region or of a nation
(macroeconomics). The following questions may serve to illustrate this
point.

While narrow economic criteria may help in answering such questions as:

o   Is the project properly scaled in size (i.e. too big, too
     small, about right)?,
o   Is the proposed timing about right? or,

ɔ   Are the design standards and components appropriate for the
     economy, or do they require excessive costs or inputs of
     human or natural resources?

Macroeconomic objectives have to be considered to respond to other
questions, such as:

ɔ   Does the sector (agriculture, power, municipal water supply,
     etc.) need additional investment?
ɔ   Does the project address the most urgent needs of this sector
     (or these sectors)? or,
ɔ   If the project does not go forward, can the resources be put
     to better uses in the country?

Another hierarchy may be established for important questions related to
each of the economic sectors included in a multi-purpose project. For
example, for the irrigated agriculture sector, the following questions
are pertinent:

At the Global Level

o   Would project output contribute towards meeting food and
     fiber needs of the country, or perhaps cause surpluses?
     (A food, or "basic needs", objective);
o   Would the project output contribute to exports and help
     improve foreign exchange balances? (A foreign exchange
     earnings objective);
o   Into what economic income categories do the project bene-
     ficiaries fall? Are these target categories compatible with
     national or regional objectives? (A distribution objective.)

At the Sector Level

o   Could the resources needed for this project be better invested
     in rainfed agriculture? (Trade-offs within agriculture
     sector.)
o   Is the irrigated farming technology proposed achievable
     within a reasonably short time period, or should the technology
     be phased in over a longer period of time? (Trade-offs over
     time.)
o   What proportion of irrigation investments should go towards
     putting new lands under irrigation, and what proportion
     should be invested in improving, rehabilitating or reclaiming
     existing or abandoned irrigation schemes? (Trade-offs
     between increasing the capital stock and rehabilitating old
     stock.)

Most engineers would recognize that the answers to such questions
require analyses which they are not trained to do. Nevertheless, it
also should be evident that the work of the engineer is vital in providing cost as well as technical information for the analyses of these
types of questions.

Public and Private Sector Projects

Irrigation and drainage projects are constructed and operated mostly by public agencies, which supplement public treasury funds by borrowing from either public or private sources. While the material in this report primarily focuses on the appraisal of public projects, most of the principles and procedures are applicable also to private sector projects. It is worth noting that private sector projects almost always require government permission and/or clearances, which can be withheld (and sometimes are) until public benefits, or lack of significant adverse effects on the public domain, can be demonstrated.

Economic vs. Financial Analysis of Projects

While economic and financial analyses are closely related as decision-making frameworks, they have important differences. Economic tests are used to estimate total return, productivity or profitability to society as a whole, from the viewpoint of a needed investment. In pure tests of economic efficiency, it is not important as to who in society specifically contributes the resources or who receives the benefits. In recent years some economists have advocated a modification of that rigid rule, claiming that it does make a difference to society as to how much of the benefits from public investments go to people who are "poor" and how much goes to people who are "rich." These aspects are discussed later.

Financial analyses, on the other hand, are made to demonstrate the narrower viewpoints of the various categories of project participants. This requires consideration of the inflows and outflows of funds to the project or parent entity, and to the farmers, cooperatives, private companies and other participants. Financial analyses seek to measure their ability to meet their financial obligations, and when appropriate, to estimate returns to equity capital, labor, and management. They are also concerned with who provides the resources, and how much, as well as to whom and how the financial benefits accrue. Financial analyses provide important information for gaging the extent to which private incentives are available to, and can be exercised by, potential project beneficiaries. While a project may be shown to be economically attractive to the nation or region as a whole, it stands to fail if farm units are unable to provide, or help to provide, their proprietors with an adequate income.

Costs and prices used in quantifying project benefits and costs differ in several important respects when used for an economic or a financial analysis. The values used for financial benefits and costs are the market prices, including taxes and subsidies. Some of the adjustments needed to convert actual financial costs or prices of goods or services to their economic counterparts are as follows:

o   All direct and indirect taxes and other transfer payments must be deducted from project financial costs because they do not constitute direct claims on physical resources and are, therefore, not economic costs. Taxes on project output, however, should be retained on the benefit side, since they are part of the value which buyers are willing to pay.

55

o   Subsidies which reduce input costs (such as for fertilizer
    or seed) must be added to project financial costs. Subsidies
    which tend to raise prices (such as support prices to en-
    courage total production), on the other hand, should be
    deducted from market values to obtain economic prices.

o   "Shadow prices" (also referred to as opportunity costs) must
    be used for quantifying cost or benefit elements when market
    prices do not reflect the real scarcities or surpluses in the
    economy. Shadow pricing of labor and of farm output are very
    common in irrigation project analyses. For some projects,
    shadow pricing of the foreign exchange rate is required when
    the project country's currency is significantly over or
    undervalued. Consideration of the shadow price of capital is
    also required in economic analyses.

o   Financial land costs must be converted to true economic land
    values. There are several alternative ways of doing this,
    one of the more common being to value the land at an estimate
    of the net value of production foregone.

o   Interest charges must be handled differently. In economic
    analyses they normally are not added to costs nor deducted
    from benefits. Some agencies add interest charges during
    construction as project costs. An economist should be con-
    sulted as to whether or not, and how, this procedure can be
    utilized in specific cases.

Situations to be Compared

Projects should be judged by comparing two situations---the way the
area would develop and produce "without" the project and the way it
would develop and produce "with" the project. An error to be avoided
is that of comparing "before" and "after" a project, for the "before"
situation in many instances could improve even without the proposed
project investments. This is particularly true in the case of an area
with existing rainfed agriculture. In the case of an arid desert area
with no existing development, the "without project" condition might be
identical to the "before project" condition, but this would be an
exception to the general rule. The quantified differences between
"without project" and "with project" conditions are referred to as
"incremental costs" or "incremental benefits."

Values of Costs and Benefits

Monetary values can be developed for most project costs and benefits,
although techniques for doing this for some of the more esoteric water
uses, such as recreation and fisheries enhancement, are still somewhat
controversial and not as well accepted as methods for establishing the
values of agricultural, power, municipal and industrial (M&I) water
supply, flood control, and navigation benefits. Perhaps the agricultural
benefits are the least controversial in methodology, but this does not
mean that they are necessarily more accurate. In special cases where
benefits are difficult to measure, but where their substantial contributions to the economy are almost without question (M&I water for a
desert community), project formulation may rely on a search for the
lowest-cost alternative for providing the same benefits. In most cases

major costs can be assigned monetary values. Where important non-quantifiable benefits or costs occur, these should be presented in special statements. Economic and financial analyses do not provide final answers, even though they are very important and influential tools to help decision makers. Intangible costs and benefits (those that cannot be quantified) associated with a project may tip the scales one way or the other in the judgment process, especially where economic and financial feasibility may be marginal.

Time Preferences

Costs and benefits of projects accrue over a period of years and include both capital investment and annual costs. Their monetary values cannot be used directly without taking into account the time value of money, i.e., $100 occurring as a benefit 10 years after the project starts has a different economic and financial value than a $100 benefit realized 5 years earlier. Also, a $100 cost incurred during the first project year has a different economic and financial value than a $100 cost incurred in later years. The differences in values in all such cases depends upon the interest (discount) rate, which represents the return to the economy (or lender, in the case of financial analyses) from the use of money. At a 6% interest rate, for example, $100 to be paid 10 years hence has a present value or present worth today of only $55.84. The rule is: the farther in time from commencement of a project, the less the present value of the future receipt or expenditure. Another rule is: the higher the discount rate, the lower the present value of the future receipt or expenditure. These relationships, coming under the heading of "discounted values of future events," are discussed in much more detail in almost all elementary economic textbooks, and in several of the references at the end of this chapter.

It is necessary that project costs and benefits be discounted properly by years before combining and/or comparing takes place. This can be conveniently done by listing the costs and benefits in vertical columns opposite the project year in which they occur. Table 5-1 is an abbreviated example, in which "without project" costs and benefits have been subtracted from "with project" costs and benefits to obtain incremental costs and benefits for each year. Costs set forth in columns are referred to as "cost streams" (cols. 3-7). Sometimes these vertical columns are loosely referred to as "cash flows," although the values may not represent cash at all. The interest or discount rate to be applied to the cost and benefit streams depends upon comparative economic parameters of the region or nation but generally should reflect the opportunity cost of capital (OCC) for investment in the project instead of investing in another project, perhaps even in a different sector of the economy. When the figures in a stream are multiplied by their appropriate discount factors, the resulting column is known as a "discounted stream." The sums of all discounted values in a column represents the present values of the cost or benefit streams. The total discounted value of the net benefits stream yields the Net Present Value (NPV).

TABLE 5-1

INTEGRATED AGRICULTURAL DEVELOPMENT PROJECT

Irrigation Component

Incremental Costs and Benefits

(in appropriate monetary units)

(1)	(2)	(3)	(4)	(5)	(6)	(7)	(8)
	Total	Incremental Costs					
	Incremental	Irrigation		Agricultural Service System		Total Incremental	Incremental Net
Year	Benefits	Construction	O&M	Construction	Operation	Costs	Benefits
1	-39	5,879	0	272	280	6,431	- 6,470
2	736	12,889	91	583	473	14,036	-13,300
3	3,053	14,145	271	1,158	583	16,157	-13,104
4	6,575	15,155	450	545	586	16,736	-10,161
5	11,482	13,771	631	545	586	15,533	- 4,051
6	15,340	831	714	0	586	2,113	13,227
7	17,257	0	714	0	586	1,300	15,957
8	18,364	0	714	0	586	1,300	17,064
9	18,549	0	714	0	586	1,300	17,249
10-25	18,882	0	714	0	586	1,300	17,582

"Incremental" refers to difference between "with" and "without" project conditions.

Specific Economic Tests

The three leading economic tests are:

(1) The Benefit/Cost Ratio (B/C ratio), which is the ratio of the present value of incremental benefits to the present value of incremental costs, based on a specified opportunity cost of capital (discount rate) and assumed project life. Over the last several years evaluation of Federal projects in the U.S.A. have been based on discount rates representing the estimated average cost of Federal borrowing as determined by the Secretary of the Treasury. At the appropriate discount rate, the B/C ratio must be at least equal to 1.0 and be higher than, or at least as high as, the B/C ratio of mutually exclusive project alternatives. (Mutually exclusive means that if one is chosen, the other cannot be undertaken).

(2) The Internal Rate of Return (IRR), which is the discount rate which equalizes the present values of the incremental benefit and cost streams over the life of the project (which also is the same discount rate which would make the Net Present Value of the Incremental Net Benefits = zero). It must equal or exceed the opportunity cost of capital.

(3) The Net Present Value (NPV), which is the net present value of the incremental benefits less the net present value of the incremental costs. Its value is determined using the same present values and costs as in the B/C ratio calculation, but the result is expressed as a difference instead of a ratio. It must not be negative at the appropriate discount rate; must be higher than, or at least as high as, the NPV of mutually exclusive alternative projects; and should be optimized.

For the project values summarized in Table 5-1 and using a discount rate of 12%, the present value of the benefit and costs streams are:

Benefits = 90,589   Costs = 53,803

The NPV would thus be 36,786. The three economic tests yield the following:

(1) B/C ratio for i = 12% = 1.68
(2) IRR   = 21.77%
(3) NPV for i = 12% = 36,786

Figure 5-1 shows the range of NPV values for various discount rates. It can be seen that for all discount rates less than the IRR, the benefits exceed costs, indicating that the B/C ratio increases for decreasing discount rates beginning with a B/C ratio = 1.0 for the IRR. For discount rates greater than the IRR, the B/C ratios are negative (insofar as can be seen from this figure). Until the advent of the modern electronic computer, calculation of discounted cash flows was a laborious time-consuming process, especially the calculation of the

59

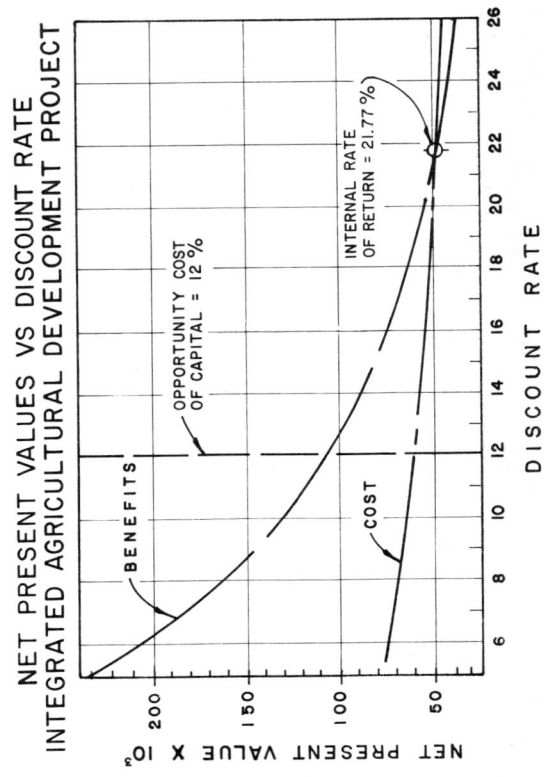

FIGURE 5-1

NET PRESENT VALUES VS DISCOUNT RATE
INTEGRATED AGRICULTURAL DEVELOPMENT PROJECT

IRR, since this requires trial-and-error iterative calculations until the correct IRR is found which makes the NPV of the net benefits equal to zero. Computers now can give results for all three tests quite readily.

For many projects and project situations, any of the three tests, properly used, can lead to a correct choice of the most economical project. For many years in the U.S.A. the B/C ratio has been the preferred index for water resources projects. Many international financial institutions make widespread use of the IRR to a considerable extent. However, it is generally agreed by economists that the NPV test should be preferred, as it is correct under all situations, while there are economic situations where the other two tests, unless special analytical precautions are taken, may not give optimum results. (Reference 8 discusses several such situations.) The World Bank presently requires its use in the analysis of water resources projects. It also is noted that the U.S. Water Resources Council, in Procedures for Evaluation of Benefits, December 14, 1979, calls for use of the maximization of net discounted benefits in scaling the level of development. This is a proper use of the NPV test.

The limitations of the B/C ratio and IRR tests are not obvious but can be learned by study of some of the references to this paper or other appropriate texts. Reference 7 (the old "Green Book" of the Federal Inter-Agency River Basin Committee) contains a good example demonstrating that the maximization of the B/C ratio would not necessarily yield an optimum project. (See also page 47.)

Another limitation of the B/C ratio method is that the ratio is sensitive to the way benefits and costs are classified, and there are no fixed universal rules in this respect, e.g., treating a cost as a negative benefit changes the B/C ratio.

Reference 8 presents several of the more important limitations of the IRR test. Problems may arise with the IRR test in some investment situations where there may be two or more IRR's which reduce the net benefits stream to zero, i.e., there is no unique solution. In such cases, one must turn to an NPV analysis to establish which would be the better investment. The B/C ratio is probably a better test for projects where operating costs are high in relation to capital costs, such as road maintenance projects, where such funds must come from limited Government budgets and where substantial benefits start early in the project life. In such cases the IRR tends to give an exaggerated impression of a project's feasibility. Indeed, the validity of the IRR becomes weaker as its value rises above the opportunity cost of capital, since the assumption that annual surpluses can be reinvested at high interest rates (implicit in the IRR calculation) becomes less and less realistic. Still, the IRR test has some advantages over the other methods, although it must yield to the NPV test for economic precision as to "go or no-go." One of its major advantages is that it does not require the pre-selection of a discount rate. Another advantage is that it permits project rankings, which the NPV test may not easily do in comparing large marginally acceptable projects with smaller highly attractive projects. It also seems to be more easily understood by laymen and financial people, and avoids, at least during the project reporting stage, arguments as to just what discount rate should be used for either the B/C ratio or NPV methods. For maximum economic correctness and presentational effectiveness, both the NPV and the IRR tests should be reported.

In calculating B/C ratios or maximizing the NPV of net benefits, the proper discount rate to use is the opportunity cost of capital (OCC). The OCC is the lowest acceptable return which capital should be expected to earn in a given country. It is represented by the earning power of capital in the marginal or last-included project in a national optimum investment program. Thus, it reflects both the supply of investment resources and the investment opportunities assumed to be available. If capital is committed to projects which cannot meet the minimum standard set by the OCC, the economy suffers from the better opportunity that is foregone. The OCC should include some premium for risk. In practice the OCC is not a precise figure. It should be set by economists and bankers rather than by water resource planners.

In most developing countries, the OCC will fall within a range of from approximately 8 to 14%; in developed countries, it will be less. For federal projects in the U.S.A., the rate used is not necessarily the OCC, but is a rate established by statute and currently is around 8%. It sometimes comes as a shock to developed-country engineers, accustomed to project design under relatively low interest rates, to learn that low-capital-cost, high maintenance-cost facilities are often the types necessary to provide economically justified projects in developing countries. For example, if the OCC is 15%, it is more economical to build and rebuild a facility costing $658 every 10 years than a single facility costing only $1,000 which will last 50 years, assuming maintenance costs of 4% of cost per year for the low-cost facility and 1% of cost per year for the high-cost facility.

Least Cost Analysis

Least cost, also called minimum-cost, tests consist of a series of present value cost estimates of a number of alternative project schemes for providing the same benefits. The alternatives usually involve different phasing of project facilities or construction, or alternative technologies or design standards. Such tests help insure that the final plan formulated will be cost-effective. All projects should be subjected to such tests. They may be the only tests possible or needed where project benefits cannot be easily or reliably quantified in monetary terms, but where the resulting benefits are considered fully justified. In water resources work they are used for sub-analyses of project components, such as water source development and transmission, and energy generation and transmission. They are often the exclusive tests for some public utilities and transportation projects.

Economic Analysis for Multi-Purpose Projects

Special issues arise in economic analyses when the costs of some facilities are shared by two or more components of a project, such as a project with both irrigation and hydroelectric power facilities. It is emphasized that these are not problems of cost allocation, which are related to financial analyses only, but problems in identifying the

most economical combination of single purpose or multipurpose projects.
Some hypothetical examples drawn from Appendix B of reference 9 [1/],
illustrate the correct application of the NPV method. In these examples,
a choice is to be made between: simple dam irrigation (e.g., a storage
reservoir plus a gravity distribution system), simple hydroelectric
scheme, tubewell irrigation, thermal power, and a multipurpose project
combining dam irrigation with hydropower and costing less than these
two projects separately. Thermal and hydropower, and dam and tubewell
irrigation, are to be mutually exclusive alternatives. It is assumed
further that the irrigation and power components for each alternative
have been properly sized, technically and economically.

TABLE 5-2

Application of Net Present Value Method

Project	Investment Costs (discounted)	Benefits (discounted)	Net Present Value
Dam Irrigation	100	90	-10
Hydropower	80	86	6
Multipurpose (dam irrigation & hydro)	150	176	26
Tubewells (Groundwater)	50	58	3
Thermal power	40	54	14
(Tubewell + thermal)	90	112	22

Note: Benefits are taken net of current costs

In the example, tubewell irrigation is preferable to single dam irrigation
and thermal power is superior to simple hydropower, but the multipurpose dam irrigation & hydro project has a higher NPV than the next
highest combination of tubewell irrigation and thermal power.

To help avoid making mistakes in selecting the best project(s), a
format such as Table 5-3 can be quite useful. The data is the same as
for the example of Table 5-2, but only NPVs are listed:

TABLE 5-3

Application of Net Present Value Method

Type Project	Irrigation	Power	Best Combination or Single Project
Single Purpose	Dam -10	Hydro 6	
	Tubewell 8	Thermal 14	22
Multipurpose			26

1/ With the permission of The World Bank.

By underlining the higher NPV in each sector and adding the two values, the best combination of single purpose projects will be established and its NPV is then inserted in the last column. If it is less than the NPV of the multipurpose project, which it is in this case, then the latter is to be preferred. If it is greater, then two single purpose projects are to be preferred.

It may happen that one single purpose project will have a higher NPV than any combination of single purpose projects. This will occur if both single purpose projects in one sector have a negative NPV, Table 5-4:

TABLE 5-4

Application of Net Present Value Method

Type of Project	Irrigation		Power		Best Combination or Single Project
Single purpose	Dam	-10	Hydro	6	14
	Tubewell	-18	Thermal	<u>14</u>	
Multipurpose					13

In this case it is the best single purpose project which is to be compared to the multipurpose project and, if it is of higher net present value, to be preferred. If no combination or project has at least zero net present value, none qualifies as an economically desirable project.

Sometimes it is not necessary to determine the cost and benefits of all four separate single purpose projects. See Table 5-5. The NPV of the hydropower project has not been established separately, but a solution still is possible. For irrigation, the NPV of the dam is greater than tubewell project and is to be preferred. The two possible combinations remaining are: 1) irrigation dam plus a thermal power project, and 2) irrigation dam plus a separate hydropower dam. The last combination of two separate dams almost always would be inferior to a multipurpose project and, hence, the only comparison to be made is between an irrigation dam plus thermal vs. the multipurpose project. In this case the dam plus thermal is better.

TABLE 5-5

Application of Net Present Value Method

Type of Project	Irrigation			Power		Best Combination or Single Project
Single purpose	Dam	<u>16</u>	(10)	Hydro	?	23 or
	Tubewells	<u>10</u>	(16)	Thermal	7	?
Multipurpose						21

64

If the figures in parantheses apply, the tubewells would be preferable
to an irrigation dam and the preceding argument would not hold. The
comparison would be between 1) tubewells plus unknown hydro and 2)
tubewells plus thermal, which has an NPV of 23 which is greater than
the multipurpose NPV of 21. The NPV of the multipurpose project can
reasonably be expected to be greater than a combination of a dam for
irrigation and a dam for power, otherwise, why have a multipurpose
alternative. Hence, it is known that the NPV of hydro alone cannot
exceed 21-10=11. However, if it is greater than 7, then the combination
of tubewells plus hydro would be better than the multipurpose project.
No choice, therefore, can be made between 1) and 2) without evaluation
of the separate hydro project.

In many situations involving power development, it is assumed that the
generation of each alternative will be the same and the levels of gross
benefit will be the same. The comparison of the hydro and thermal
options is thus reduced to a cost comparison rather than a comparison
of costs and benefits. This causes some analytical complication as
illustrated by the example of Tables 5-6 and 5-7. The NPV of the power
produced, which is the same for both thermal and hydro, is designated
by "e," which is related to the economic value of power to the country.

TABLE 5-6

Application of Net Present Value Method

Project	Investment Costs (discounted)	Benefits (discounted)	Net Present Value
Dam irrigation	100	113	13
Hydropower	100	e-113	e-113
Multipurpose	150	e+100	e-50
Tubewell irrigation	50	68	18
Thermal power	40	e-54	e-94

TABLE 5-7

Application of Net Present Value Method

Type of Project	Irrigation		Power		Best Combination or Single Project
Single purpose	Dam	13	Hydro	e-113	e-76, (-94+18)
	Tubewell	18	Thermal	e- 94	or 18 if e<94
Multipurpose					e-50

The multipurpose project is the best choice of the alternatives which
include power. Tubewells are the best single purpose project. The
multipurpose project has a higher net present value only if e-50>18.
If e<68, tubewell irrigation by itself is the best choice. Thus, the
final selection depends upon the value of "e". The general conclusion
is that the value attributed to the power benefits may determine

65

whether the multipurpose project should be carried out. It is possible
without knowing the power benefits to determine the best project among
the single purpose power projects, the combinations of irrigation and
power projects, and the multipurpose project, for in these comparisons
"e" drops out. But since the multipurpose net benefits are only expressed
in a form such as "e-50", it is uncertain that they are positive, or
that they are greater than the NPV from the better single purpose
project. However, to determine this it will not always be necessary to
estimate the value of "e" precisely, but only to say whether it is
sensibly above or below the relevant figure---in this case, 68.

Project Life

The appropriate time period for project life is the economic life of
the major investment, rather than the technical life. For tubewell
projects a life of 15-25 years might be appropriate, while for a gravity
irrigation project, with many canal networks, a life of from 50-100
years is technically and economically acceptable. With the aid of
electronic computers, it is not difficult to incorporate the benefits
and costs over as long a life period as desired. However, it is worth
noting that benefits or costs occurring in later years may have negligible
affects on the analysis because of the discount factor. Table 5-8
shows the present values of a $100 benefit or cost occurring in the
years indicated:

TABLE 5-8

Present Value of $100 at a Future Date

Project Year	Discount Rate			
	6%	10%	15%	20%
20	31.18	14.86	6.11	2.61
25	23.30	9.23	3.04	1.05
30	17.41	5.73	1.51	0.42
40	9.72	2.21	0.37	0.07
50	5.43	0.85	0.09	0.01
60	3.03	0.33	0.02	0

Irrigation projects in developing countries normally must return 10-15%
to meet OCC requirements. Hence, periods of analysis for such projects
can be appreciably less than projects in developed countries where the
OCC may be 6% or less. If a client in a developing country insists on
longer periods of analysis because the dam may be expected to remain
for 50, 100 or more years, there is little to be gained by prolonged
arguments as the economic results will be unaffected.

66

Indirect or Secondary Benefits and Costs

Many financial agencies do not permit the use of indirect or secondary
benefits in economic or financial analyses, usually on the basis that
determination of such benefits is still highly subjective in many
instances. Many do, however, request that such benefits be at least
described. Other agencies permit, or even request, that such benefits
be included. In any event, if such benefits are included, costs associated
with the realization of such benefits also should be identified and
included.

Social Analysis

For some time, with increasing intensity, arguments have been advanced
that traditional economic analyses--which emphasize economic efficiency--
should be modified to reflect social preferences for projects benefitting
designated target groups, (e.g., for agricultural projects, the rural
poor). One simple way of doing this would be to lower the required
opportunity cost of capital for projects with such emphasis. But how
much should it be lowered? And since it is difficult in many cases to
find project areas where all are poor and where none have above-average
incomes, how should such mixed cases be treated? One of the more
complete discussions of a systematic approach to such cases is given in
Reference 3, but it is written for professional economists; not for
other professionals or laymen. The approach is to modify net benefits
to "net social benefits" in accordance with complex calculations involving
numerous national economic and social parameters. One of the concepts
set forth therein is that of the Critical Consumption Level (CCL). At
this level of per capita income, the real resource cost incurred by the
Government by project investment and the social benefit enjoyed by the
worker as a result of marginal increase in consumption are exactly
offsetting. At this level, the social price equals the efficiency
price. If a government wishes to give preference to people with incomes
below that level, the methodology gives more weight to benefits going
to that group. It can also give decreased weight to benefits going to
those with incomes above the CCL, if desired. The CCL varies considerably
from country to country; as a rough guide, it is likely to range between
one third and two thirds of the national average-level of income. The
CCL has some importance in project cost-recovery considerations in
projects dealing with small farmers since, under this theory, the
Government (society as a whole) gains little or nothing by requiring
people with incomes below the CCL to repay project costs. If it does
so, it may have to give the money back to them for other purposes.

Financial Aspects

General Considerations

Obviously a project cannot be built unless funds (resources) are assured
from one or more sources for this purpose, on a timely basis and on
reasonable terms. This does not mean that construction fund pledges
are needed to cover the ultimate costs of a large project, but it does
mean that sufficient funds should be available to enable timely completion
of economically and financially feasible phases of the project. Such

funds may come from internal or external sources, or both. For irrigation projects it is very rare that all funds would be available from outside sources. Seldom does the project formulation process require that the viability of the outside financial sources be checked but, occasionally, this could occur, as for example, when small new banks may offer to help finance a project. Reasonableness of the terms is important because the borrower must be able to repay the loans. The ability to do so is one of the items to be checked as part of the financial analysis. The proposed project financing plan should be presented in a summary form, such as the example in Table 5-9.

The financial analysis also should insure that there will be sufficient resources available to operate and maintain the project, and to repay the loan obligations. In the case of many irrigation projects the project entity itself may not be able to, or responsible for, loan repayment. Indeed, the actual financial performance and arrangements for many projects are so poor that O&M expenses collected from water users are insufficient, and the balance must come from public treasuries which, of course, must have collected the money by other methods. It is a common need for public irrigation or drainage projects, therefore, to have pro forma financial projections which will show annual accruals or disbursements to general public treasuries. Unfortunately, this common need is often neglected by project planners. Tables 5-10 and 5-11 are sample projected cash flow projections for a parastatal agricultural organization in a developing country.

The financial analyses for power and water supply entities which may be involved in a multipurpose project are more complex than for irrigation projects, since requirements for reserve funds, depreciation allowances, return on equity, ratios of debt to equity and earnings to future debt service, are standard considerations. Such analyses should be performed by experts in financial analysis of utility projects.

Agricultural Budgets

For an irrigation project, the development of farm-model budgets is a basic exercise which should cover representative farm types. References 2 and 5 have suggestions in this regard, but the widespread variations in farms, crops and systems mandate some flexibility in format. Table 5-12 gives basic data on both economic and financial costs used in projections for a proposed project, and Tables 5-13 through 5-15 are sample presentations.

It is extremely important that farmers in an irrigation project have financial incentives to participate, and the farm budget is the source of some guidance in this regard. Their current estimated financial income and projections, both with and without the project, usually are needed. In some cases the "present" income also can be taken as "future without," but this always should be carefully considered as it may be challenged by reviewing agencies. A common "future without" situation is slightly increased yields through existing extention service efforts and future projected crop prices. Any increased "future without" costs also should be taken into account.

TABLE 5-9

AGRICULTURAL DEVELOPMENT PROJECT

Project Financing Plan

(in thousands of monetary units)

	Int. Fund for Agr. Dev.			Int. Devel. Assoc.			Cotton Org. & Farmers			Government			Total Project Costs		
	Local Cost	Foreign Expend.	Total	Local Cost	Foreign Expend.	Total	Local Cost	Foreign Expend.	Total	Local Cost	Foreign Expend.	Total	Local Cost	Foreign Expend.	Total
Major Civil Works	-	-	-	-	990	990	-	-	-	389	-	389	389	990	1,379
Land Development															
Clearing and Levelling	-	1,058	1,058	-	-	-	-	-	-	397	-	397	397	1,058	1,455
Canals	-	-	-	413	577	990	-	-	-	368	-	368	781	577	1,358
Sub-Total	-	1,058	1,058	413	577	990	-	-	-	765	-	765	1,178	1,635	2,813
Design and Supv.	-	-	-	19	71	90	-	-	-	52	-	52	71	71	142
Inc. Production	-	-	-	-	-	-	356	426	782	-	-	-	356	426	782
Applied Research	-	1	1	-	-	-	-	-	-	3	-	3	3	1	4
Social Infrastructure	-	-	-	-	-	-	13	-	13	10	10	20	23	10	33
Technical Services															
Project Prep. (Mangoky)	-	-	-	-	236	236	-	-	-	199	60	259	199	296	495
Applied Investigation	-	21	21	-	-	-	-	-	-	17	3	20	17	24	41
Training	-	-	-	-	15	15	-	-	-	11	-	11	11	15	26
Project Prep. (MDR)	-	180	130	-	42	42	-	-	-	149	-	149	149	222	371
Sub-Total	-	201	201	-	293	293	-	-	-	376	63	439	376	557	933
Contingencies 1/	73	129	202	122	215	337	-	-	-	-	-	-	195	344	539
Total	73	1,389	1,462	554	2,146	2,700	369	426	795	1,595	73	1,668	2,591	4,034	6,625

1/ This category does not include the total amount of contingencies, but corresponds only to the unallocated parts of the IDA credit and the IFAD loan. Consequently, the costs of the item financed by IDA and IFAD have been reduced by that amount in the above table.

Adapted with the permission of The World Bank.

TABLE 5-10

AGRICULTURAL DEVELOPMENT PROJECT

Projected Cash Flow from Agricultural Activities

Description	1981	1982	1983	1984 onward
Existing EDF on-farm Development (ha)	5,400	5,400	5,400	5,400
New on-farm Development IDA-FIDA (ha)	1,800	2,550	3,300	3,300
Total Developed (ha)	7,200	7,950	8,700	8,700
Areas cultivated (ha)				
- Cotton	4,250	4,300	4,550	4,600
- Rice	1,700	2,400	2,900	3,400
- Antaka	500	500	500	700
	6,450	7,200	7,950	8,700
Income (X FMC 1000)				
- Cotton (FMC 40/kg and 2.8 t/ha)	476,000	481,600	509,540	515,120
- Rice (2 t/ha and FMC 33/kg)	119,000	168,000	203,000	238,000
- Antaka (1 t/ha and FMC 20/kg)	10,000	10,000	10,000	14,500
- 50% Rebate on chemicals on Cotton	187,000	189,200	200,200	202,200
- Rebate on spray	13,600	13,760	14,620	14,500
	805,600	862,560	937,360	984,320
Expenses (X FMC 1000)				
a) Direct Costs				
- Cotton (FMC 117,520/ha)	499,460	505,336	534,716	540,592
- Rice (FMC 5,300/ha)	77,010	108,720	131,370	154,020
- Antaka (FMC 15,650/ha)	7,825	7,825	7,825	10,955
- Network Maintenance (FMC 7,870/ha)	50,761	56,664	62,566	68,469
	635,056	678,545	736,477	774,036
b) Staff	88,050	98,700	108,200	108,200
c) Overheads	34,000	36,000	38,000	40,000
Total Expenses = a + b + c	757,106	813,245	882,677	922,236
Profits = Income - Expenses	48,494	49,315	54,683	62,084

TABLE 5-11

AGRICULTURAL DEVELOPMENT PROJECT

Projected Cash Flow

(in thousands of monetary units)

Description	1981	1982	1983	1984 onward
Total Income				
Agriculture	805,600	862,560	937,360	984,320
On-Farm Development	350,000	350,000	350,000	350,000
Canal Factory	150,000	150,000	150,000	150,000
Total Income	1,305,600	1,362,560	1,437,360	1,484,320
Expenses				
Agricultural Expenses:				
- Direct Costs	635,056	678,545	736,477	774,036
- Staff	89,050	98,700	108,200	108,200
Total	724,106	777,245	844,677	882,236
On-Farm Development Expenses				
- Personnel	90,000	90,000	90,000	90,000
- Materials	50,000	50,000	50,000	50,000
- Equipment	160,000	160,000	160,000	160,000
Total	300,000	300,000	300,000	300,000
Canal Factory Expenses				
- Personnel	40,000	40,000	40,000	40,000
- Equipment	30,000	30,000	30,000	30,000
- Materials	80,000	80,000	80,000	80,000
Total	150,000	150,000	150,000	150,000
Overheads	81,000	83,000	85,000	87,000
Total Expenses	1,255,106	1,310,245	1,379,677	1,419,236
Profits	50,494	52,315	57,683	65,084

Adapted with the permission of The World Bank.

TABLE 5-12

AGRICULTURAL DEVELOPMENT PROJECT

Prices of Inputs and Outputs

	Present		Future [1]	
	Financial	Economic	Financial	Economic
Crops				
Cotton (per ton)	83,000	77,130	83,000	82,840
Rice (per ton)	35,000	48,170	35,000	60,990
Fertilizer				
Urae (per ton)	45,000	45,000	50,600	50,600
Pesticides				
#1	800,000	800,000	800,000	800,000
#2	2,000,000	2,000,000	2,000,000	2,000,000
#3	3,000,000	3,000,000	3,000,000	3,000,000
#4	450,000	450,000	450,000	450,000
Seeds				
Cotton (per ton)	10,000	10,000	10,000	10,000
Rice (per ton)	40,000	40,000	40,000	40,000
Custom Services				
Cotton	54,780	54,780	54,780	54,780
Rice	21,860	21,860	21,860	21,860
Water	10,000	10,000	10,000	10,000

[1] Full development (after Project Year 12).

Adapted with the permission of The World Bank.

TABLE 5-13

AGRICULTURAL DEVELOPMENT PROJECT

Financial Crop Budgets for Farm Models

Crop[1]	Yield[2] (tons/ha)	Actual[2] Yield (tons)	Price	Gross Value	Seed	Ferti- lizer	Agroche- micals	Other	Total Costs	Net Value	Labor Require- ments (man-days)
						In 1,000s					
Present											
Model 4 Rice	1.8	1.8	35	63	1	0	0	0	1	62	300
Model 5 Rice	1.8	1.8	35	63	1	0	0	0	1	62	300
Future w/ Project											
Model 1 Rice	7.0	7.0	35	245	2	10	3	55	70	175	300
Model 2 Rice	7.0	7.0	35	245	2	10	3	55	70	175	300
Cotton	2.8	1.4	85	119	0	8	37	11	56	63	100
Model 3 Rice	7.0	3.5	35	123	1	5	1	27	35	88	200
Cotton	2.8	2.8	85	238	1	15	75	22	112	126	200
Model 4 Rice	2.6	2.6	35	91	1	0	0	10	11	80	300
Model 5 Rice	6.4	6.4	35	224	2	10	3	55	70	154	300

1/ Models 1, 2, 3 represent farms in the extension areas. Therefore, no "present" or "future without project" situation is given. Models 4 and 5 represent farms in the Mangolovolo area. For those, the "future without project" situation is assumed to be equal to the "present" situation.

2/ Yields for "future with project" are yields at full development (after Project Year 12). Rice yields two crops per year.

Adapted with the permission of The World Bank.

73

TABLE 5-14

AGRICULTURAL DEVELOPMENT PROJECT

Operational and Financial Results

Cash Flow (Farm 1)
(In 1,000s)

Year	1	2	3	4	5	6	7	8	9	10	11	12	13	14	15
Gross Value of Production	211	223	245	245	245	245	245	245	245	245	245	245	245	245	245
Cost of Inputs	70	70	70	70	70	70	70	70	70	70	70	70	70	70	70
Hired Labor	1	1	1	1	1	1	1	1	1	1	1	1	1	1	1
Farm Income	140	152	174	174	174	174	174	174	174	174	174	174	174	174	174
Net Cash Flow															
Annual	140	152	174	174	174	174	174	174	174	174	174	174	174	174	174
Cumulative	140	292	466	639	813	987	1,161	1,335	1,509	1,683	1,857	2,030	2,204	2,378	2,552

Adapted with the permission of The World Bank.

TABLE 5-15

AGRICULTURAL DEVELOPMENT PROJECT/FINANCIAL RETURNS TO LABOR

	Gross Value (units)	Cost Inputs (units)	Net Value (units)	Labor Req. (man-days)	Return to Labor [1] (units/man-days)
Model 1 (No. & West. Ext.)					
1.0 ha of rice	245,000	70,000	175,000	300	600
Model 2 (No. & West. Ext.)					
1.0 ha of rice	245,000	70,000	175,000	300	600
0.5 ha of cotton	118,000	56,000	63,000	100	700
Total	364,000	126,000	238,000	400	600
Model 3 (No. & West. Ext.)					
0.5 ha of rice	122,500	35,000	87,500	200	600
1.0 ha of cotton	238,000	112,000	126,000	200	700
Total	360,500	147,000	213,500	300	700
Model 4 (Mangolovolo)					
1.0 ha of rice	91,000	11,200	79,800	300	300
Model 5 (Mangolovolo)					
1.0 ha of rice	224,000	70,000	154,000	300	600

[1] The difference in returns to labor between rice and cotton cultivation reflect input subsidies for cotton production. Without these subsidies, returns to labor would be higher in rice cultivation than in cotton cultivation.

Adapted with the permission of The World Bank.

Clear identification of proposed water charges as a separate line cost item is very helpful in preparing farm budgets. It can then be easily omitted as a deduction to help establish "ability to pay" water charges. This will be discussed later.

Financial statements showing earnings, annual revenues, and expenses, return to equity, and pro forma balance sheets are often required for private agricultural enterprises. Three examples are shown in Tables 5-16, 5-17, and 5-18.

Cost Recovery

Background

Cost recovery is one of the most, if not the most, controversial aspect to be encountered in formulation of an irrigation and drainage project and during its operation. While charges for electric power and municipal and industrial water supplies also can be controversial, there is much less controversy proportionally in those sectors than in the irrigation sector. The reasons for this difference are complex, but it is sufficient to recognize that the issue of cost recovery for irrigation project operation and maintenance (O&M) costs and for investment costs is one which must be faced during project formulation. The extent and manner of cost recovery directly affects the financial cash flows of the farmer, the project organization, and more than one government organization as well as affecting the national income, and therefore, must be realistically approached. Cost recovery also can affect the economics of a project. If cost recovery plans impose too heavy a potential burden on the farmers, there may be insufficient incentives for them to fully participate in the project; and the projected output will not be achieved. On the other hand, if O&M budgets are insufficiently supported by farmers and/or the Government, water deliveries to farmers may be unreliable and insufficient, and production could again suffer. As a result of this sensitivity of farmer and Government incomes to cost-recovery decisions, these are almost always political issues to some degree. Since it involves human reactions and opinions, as well as hard financial facts which ultimately have to be faced, it is not something which can be described adequately by arithmetic procedures and financial tables only, although they do give importance guidance to decision makers.

Cost recovery can be affected by both direct and indirect measures. Around the world there are wide variations on the magnitude, nature and extent of cost recovery. Annual charges of as much as $1,000/ha have been imposed for some newly-developed tidelands in the Far East, while some governments are known to charge as little as $1-2/ha/yr. More common values are in the range of from $30-150/ha/yr.

Basic Aspects

Three important basic aspects of cost-recovery analysis are:

o   Economic Efficiency: The extent to which scarce water resources are optimally allocated among different uses.

TABLE 5-16

PROJECTED PROJECT REVENUE BY QUARTER AND POTENTIAL EARNING POWER (IN $1,000)
NO LEVELING PAYMENT NOR WAIVER OF CHARGES
RETURN ON TOTAL INVESTMENT 18.37 PERCENT

Fiscal Year Quarter	Gross Farm Revenue (1)	Direct Operat. Cost (2)	Management Cost (3)	Net Farm Revenue (4)	Total Net Revenue (5)	Total Invest- ment (6)	Annual Cash Balance (7)	Cum. Cash Balance (8)
Yr 1								
2	0	0	0	0	0	-376	-376	-376
3	0	-5	-1	-6	-6	-671	-677	-1,053
4	9	-139	-25	-155	-155	-363	-518	-1,571
1	28	-90	-16	-78	-78	-275	-353	-1,924
YEARLY TOTAL	37	-234	-42	-239	-239	-1,685	-1,924	-1,924
Yr 2								
2	210	-149	-27	34	34	-264	-230	-2,154
3	208	-142	-26	40	40	-248	-208	-2,362
4	131	-320	-58	-247	-247	-217	-464	-2,826
1	85	-215	-39	-169	-169	-226	-395	-3,221
YEARLY TOTAL	634	-826	-150	-342	-342	-955	-1,297	-3,221
Yr 3								
2	542	-471	-85	-14	-14	-144	-158	-3,379
3	809	-274	-49	486	486	-34	452	-2,927
4	367	-704	-107	-444	-444	-48	-492	-3,419
1	125	-287	-107	-269	-269	-40	-309	-3,728
YEARLY TOTAL	1,843	-1,736	-348	-241	-241	-266	-507	-3,728
Yr 4								
2	1,520	-765	-107	648	648	-40	608	-3,120
3	1,352	-385	-107	860	860	-40	820	-2,300
4	506	-874	-107	-475	-475	-40	-515	-2,815
1	154	-264	-107	-217	-217	0	-217	-3,032
YEARLY TOTAL	3,532	-2,288	-428	816	816	-120	696	-3,032

TABLE 5-17

PROJECTED OPERATING STATEMENT ($1,000)

NO LEVELING PAYMENT NOR WAIVER OF CHARGES
MACHINERY FINANCING 5 YEARS, 25 PERCENT EQUITY, 8 PERCENT INTEREST

Year Ending March 31:

	1	2	3	4	5	6	7	8
Total Sales	37	635	1,843	3,531	4,165	4,369	4,371	4,371
Other Income	0	0	0	0	0	0	0	0
Change in Inventory	38	72	246	120	0	0	0	0
Change in Planted Crops	293	362	369	45	29	-1	1	2
Total Gross Income	368	1,069	2,458	3,696	4,194	4,368	4,372	4,373
Direct Farm Expense	234	826	1,736	2,288	2,388	2,402	2,402	2,402
Management Expense	41	150	348	428	428	428	428	428
Interest	60	158	229	261	265	269	272	274
Depreciation								
Buildings & Development Cost	30	57	67	67	67	67	67	67
Vehicles	16	35	41	41	32	23	20	20
Farm Machinery (Major Supplier)	19	45	59	70	73	73	73	73
Farm Machinery (Other Supplier)	1	5	7	11	11	12	15	17
Total Depreciation	66	142	174	189	183	175	175	177
Total Expense	401	1,276	2,487	3,166	3,264	3,274	3,277	3,281
Net Income Before Tax	-33	-207	-29	530	930	1,094	1,095	1,092
Income Tax	0	0	0	0	0	0	310	309
Net Income	-33	-207	-29	530	930	1,094	785	783

TABLE 5-18

PROFORMA BALANCE SHEET ($1,000)

NO LEVELING PAYMENT NOR WAIVER OF CHARGES

MACHINERY FINANCING 5 YEARS, 25 PERCENT EQUITY, 8 PERCENT INTEREST

Year Ending March 31

	1	2	3	4	5	6	7	8
Current Assets:								
Cash	115	138	205	730	1,798	3,054	4,005	4,955
Livestock Inventory	38	110	356	476	476	476	476	476
Equipment Inventory	994	1,283	1,276	1,396	1,456	1,521	1,553	1,585
Less Depreciation	-142	-346	-557	-678	-794	-902	-1,009	-1,119
Net Inventory Equipment	852	937	719	718	662	619	544	466
Growing Crops	293	655	1,024	1,068	1,098	1,095	1,096	1,098
Total Current Assets	1,298	1,840	2,304	2,992	4,034	5,244	6,121	6,995
Fixed Assets:								
Investment in Development	805	1,592	1,845	1,845	1,845	1,845	1,845	1,845
Less Depreciation	-30	-87	-154	-221	-289	-356	-423	-490
Net Investment in Development	775	1,505	1,691	1,624	1,556	1,489	1,422	1,355
Total Fixed Assets	775	1,505	1,691	1,624	1,556	1,489	1,422	1,355
Total Assets	2,073	3,345	3,996	4,616	5,590	6,733	7,543	8,350

(Continued On Next Page)

TABLE 5 18

(Continued)

Current Liabilities								
Accounts Payable, Chemicals	66	227	372	418	430	430	430	430
Accounts Payable, Fuels	1	23	39	63	73	73	73	73
Accounts Payable, Water	27	38	55	57	61	61	61	61
Accounts Payable, Other	30	47	79	87	88	88	88	88
Production Loan	96	262	602	741	754	755	755	754
Machinery Notes, Current	78	88	87	76	48	21	9	3
Chattel Mortgages, Current	95	87	20	-3	7	23	19	15
Total Current Liabilities	393	772	1,254	1,439	1,461	1,451	1,435	1,424
Deferred Liabilities								
Machinery Notes, Deferred	314	353	348	303	194	84	38	11
Chattel Mortgages, Deferred	188	173	40	-8	14	46	38	30
Development Lease Loan	685	1,391	1,557	1,555	1,665	1,802	1,896	1,966
Total Deferred Liabilities	1,187	1,917	1,945	1,850	1,873	1,932	1,972	2,007
Total Liabilities	1,580	2,689	3,199	3,289	3,334	3,383	3,407	3,431
Net Worth								
Paid-In Capital	525	896	1,066	1,066	1,066	1,066	1,066	1,066
Capital Reserve	0	0	0	30	45	61	69	77
Earned Surplus	-32	-239	-269	231	1,145	2,223	3,001	3,776
Total Net Worth	493	657	797	1,327	2,256	3,350	4,136	4,919
Total Liabilities & Net Worth	2,073	3,346	3,996	4,616	5,590	6,733	7,543	8,350

o   **Income Distribution**: The manner in which the benefits flowing
     from irrigation are shared among project beneficiaries; and

o   **Public Savings**: The extent to which Government captures part
     of the increased net benefits for funding future investment
     in agriculture and elsewhere.

Economic Efficiency

This aspect is concerned with maximizing a project's net benefit to the
economy. True efficiency pricing is rarely encountered in irrigation
projects, since it requires volumetric measurement of water deliveries
(or accurate estimates), plus a sales mechanism to obtain market-
clearing prices for water. There are both technical and political
problems in such requirements. However, water should never be offered
as a "free" good. Even a nominal price for water offers some incentive
to reduce or minimize wastage, and metering has been demonstrated in
most instances to reduce usage. Metering means extra costs, but it
should not be automatically ruled out, especially where the costs of
supplying the water are relatively high. Direct pricing of water at
near-efficiency rates may conflict with the income distribution goals,
and hence, other methods of assessing charges may also have to be
considered to meet those goals and ensure that charges would be within
the capacity of beneficiaries to pay and would leave them with adequate
incentives to participate. Also, charges should not be so high as to
discourage farmer participation in a project or to tend to encourage
under-irrigation and salinity buildup.

Income Distribution

Some governments (including the U.S.A.) have policies which call for
increasing the income of the poorer portions of society from publicly-
financed projects or limiting to some degree potential unearned profits
of the affluent beneficiaries. This automatically means a reduced
share of incremental benefits to those with relatively richer income
levels. Acreage limitation is just one way used historically in the
U.S. for this purpose. Graduated income taxes are another (Graduated
taxes are sometimes called "progressive"). There seems to be a great
psychological barrier to the use of progressive irrigation water charges
where larger landowners would pay more per unit of water or area than
smaller landowners. It is argued that water should be viewed as other
farm inputs, such as seed and chemicals, and that all should pay the
same unit price per product. More recently, there has been limited
acceptance of progressive electricity charges in the U.S.A.; but by and
large, progressive, direct irrigation water charges are still only a
possibility in most of the world.

Uniform charges per unit of water or land frequently can result in the
total water charges, expressed as a percentage of the increased net
income due to irrigation, becoming increasingly greater as the size of
the farm decreases, especially as sizes drop down to around ten hectares
or less. Such a result is termed "regressive", and generally, if
possible, should be avoided entirely or, at least, minimized.

For farmers with incomes below the Critical Consumption Level (CCL)
there is economic justification for no imposition of water charges. In
principle, those who remain below the CCL are judged to be so poor that
no additional tax burden should be imposed on them while those above
the CCL should be taxed progressively by a proportion of their increased
benefits. However, primarily for economic and water efficiency reasons,
even people at these lower-level incomes should be expected to pay
something for irrigation water service.

Public Savings

Farmers who receive the benefits from public investments in irrigated
agriculture are immediately transferred into an "elite" privileged
group compared to farmers who could benefit from, but do not have, such
facilities. Those benefitting should be expected to return part of
their increased incomes to Government to help build future projects for
others or for other desirable public investments. All governments have
limited resources; and unless investment projects return extra revenue
to the treasury directly or indirectly, development cannot proceed very
far.

Costs of Operation and Maintenance (O&M)

Unless sufficient funds are collected and used for annual O&M expenses,
project facilities will gradually deteriorate; and the system will be
unable to provide timely and adequate irrigation supplies. The funds
must come from the beneficiaries or Government, or a combination of
both. It has been observed that, in general, where systems receive all
their O&M funds from the beneficiaries and the beneficiaries have a
significant influence on the operation of the system, most such systems
are fairly well-maintained. On the other hand, where water supplies
are inadequate or untimely, farmers are reluctant or outright refuse to
pay for poor services. In some places, farmers' payments go to a
treasury other than that of the operating agency; and the agency and
farmers are dependent upon Government disbursements for O&M. In other
instances, local farmer organizations do a good job of providing O&M
for small areas of systems; then again, sometimes in the same country
similar local organizations are ineffective.

The reasons for these differences are complex and may not necessarily
be resolved by a frequently advocated rule: "As a minimum, water
charges should cover O&M costs." Adoption of such a policy certainly
is a firm first step in meeting economic efficiency and public savings
goals. It also would mean that projects would not be an annual drain
on Government budgets, provided that the O&M costs represented adequate
O&M. Unfortunately, this frequently fails to be the case.

Furthermore, payments limited to O&M costs generally should be viewed
as only an initial step which should be augmented by increased charges
to help recover some of the public investment costs and meet public
savings objectives. Depending upon the level of O&M charges, they may
or may not be consistent with income distribution goals, but usually,
O&M charges are not greatly out-of-line with this objective.

Summary Statement of Principles

Recovery of irrigation project costs by direct and/or indirect methods
should cover O&M costs, plus as much as reasonably possible of capital
costs, taking into consideration the capacity of beneficiaries to pay,
the need for them to have adequate incentives to participate in the
project, possible disincentives, problems of tax evasion and cost
collection, and differences in income levels.

Repayment Capacity, or Project Rent

An important step in developing cost-recovery policies and proposed
levels is an analysis of farmer incomes for various representative farm
sizes included in a project. Such an analysis begins with preparation
of a financial farm budget which is determined as follows:

Start With:

Incremental Gross Value of Farm Production-

Subtract:

o   *Incremental Cash Farm Expenditures,* except *for irrigation
      water charges or taxes,*

o   *Incremental Depreciation of Farm Assets,*

o   *Incremental Imputed Values for Farm Family Labor and Management,*

o   *Allowance for Return on Incremental Farmer's Own Capital,*

o   *Incremental General Taxes,*

o   *Allowance for Additional Risk/Uncertainty.*

Results In:

Capacity to Pay Irrigation Water Charges.

Repayment capacity is a form of "project rent," a term favored by many
economists. Theoretically, the entire amount of rent could be recovered
from the beneficiaries by direct and indirect measures, and they would
still have adequate incentives to participate in the project. In
practice, estimates of rent can only be approximate, since several of
these elements are difficult to determine or estimate. The proportion
of the rent which can actually be recovered without undue difficulty
depends upon so many unmeasurable factors that a cautious approach is
prudent. Comparisons of estimates and recovery performances in other
instances may be helpful but cannot be completely transferred. There
is some evidence that generous reductions in the rent estimates are
necessary to bring proposed charges in line with farmers' willingness
to pay, which is almost always less than what a paper analysis might
indicate they are able to pay.

83

The word "incremental" is used to indicate that the comparison should be "with" and "without" project. An analysis is also needed of repayment capacity or rent omitting the incremental feature so that the relationship to the final farm budget can be known. Both the incremental and full figures should be made for the situation when projected yields are reached under full development conditions. This is especially important when rehabilitating an old system and extending its service to areas presently without irrigation. Here the incremental benefits are quite different for farmers in the old and new service areas, but it would be difficult to impose different water charges.

Charges to recover capital costs normally are imposed gradually, sometimes only after an initial grace period. Comparisons of the amounts required over a period of time to recover capital costs (in addition to covering all O&M costs) can be related to proposed charges. All should be put in cash-flow form and properly discounted. Proper account should be taken of interest, inflation and subsidies.

Cost Recovery Methods

Some recovery of benefits and costs normally will result from existing general taxes, such as income or export taxes. However, capturing a large part of the costs through an increase in general taxation impinges also on those who do not directly benefit from a project. Changes in general taxation should be decided on broader fiscal grounds and not within the context of a particular project. Hence, any recovery of costs through taxes should affect, as much as possible, the project beneficiaries only. Such measures are sometimes referred to as "benefit taxes." The special land taxes imposed by many western U.S.A. irrigation districts fall into that category and are designed to recover some of the benefits from those living within the district who indirectly benefit from the increased economic activity due to the project as well as from direct beneficiaries. While it is recommended that general taxes not be increased solely because of a project, it is legitimate to consider the increased revenues from such taxes resulting from increased project beneficiary incomes as part of the indirect cost recoveries for a project.

Mechanisms to recover project costs which are commonly used or are not uncommon are:

o   direct water charges per unit volume or area
o   project area assessments on land or property
o   taxes in cash or kind on project production
o   control of prices to levels below world market prices, and
o   compulsory sale of product to parastatal agencies at prices below world market prices.

While all of these are appropriate cost recovery, and indeed in some instances revenue-producing mechanisms, they can backfire if misused and destroy or lessen farmers' incentives to produce. Recovery of project costs is a very important, but very complex, process. For formulation of new projects the advice of both local and more universally experienced experts in taxation and water charges should be sought.

Cost Allocation

Need for Cost Allocation

Irrigation components are frequently important parts of multipurpose water resources development projects because the costs of some major features may be shared by more than one user, hence, reducing the costs below that of separate projects providing the same benefits. While this should be true for properly formulated projects, it is not inherently or universally true for all proposed multipurpose projects. Whenever all or portions of project costs are required to be recovered, a major problem arises as to how these joint savings should be shared. Alternatively stated: How should these joint costs be allocated and shared? Since this is a commonly encountered problem in irrigation project formulation, those doing the formulation should be familiar with some of the fundamental principles underlying cost allocation calculations.

Lack of Theoretical Basis

A key point to be understood at the outset is that there is no basis in economic theory for any known cost allocation method. This point does not seem to be made in most of the literature on the subject, and since cost allocation calculations in the majority of cases are discussed or treated under the general heading of "economic work," many non-economists may have the mistaken impression that somewhere underneath the mass of calculations used in some allocation procedures lies a theoretical foundation. Knowing that this is not the case may help in stimulating the application of the simpler types of allocation procedures appropriate to the circumstances. However, this does not mean that cost allocation methods are totally devoid of economic or financial rationality. It means that no one method is inherently better than another method from the standpoint of economic theory only.

Policies Needed

Cost allocation is an exercise in logic and equity bounded by financial, legal, administrative and political guidelines dependent upon simple but often laborious arithmetic processes, and in practice, pierced by inconsistencies and inadequacies. The policy of the U.S.A. Water Resources Council, as set forth on p. 140 of Reference 6, is a good statement for a guiding policy:

> "Reimbursement and cost-sharing policies shall be directed to the end that identifiable beneficiaries bear an equitable share of costs commensurate with beneficial effects in full cognizance of the objectives." ("objectives" refers to national water development policies.)

Such a policy statement immediately establishes some boundaries on the cost allocation process by directing that the allocation be "equitable, commensurate with beneficial effects and with cognizance of national objectives." Other obvious boundaries are ability and willingness to pay by the beneficiaries and ability and willingness to collect by the owners. No universally accepted method applicable to all situations is known.

Cost Allocation Methods

There are several methods of allocating the costs of joint-facilities. Some of the more common are:

- Proportional Use of Facilities
  on basis of volume,
  on basis of capacity, and
  on basis of volume and capacity
- Proportional Benefits must be limited by cost of viable least-cost alternatives
- Alternative Justifiable Expenditure; and
- Separable Cost-Remaining Benefits.

Separable Cost-Remaining Benefits (SCRB) Method

The SCRB method is the one found to be generally most useful in the U.S.A. and for many foreign projects which include two or more purposes such as irrigation, hydropower, water supply, flood control, and navigation. It does have weaknesses though, and often logical adjustments must be made to adopt it to special situations. It is described in some detail in References 1, 2, 6, and 7. Reference 6 sets forth modified procedures for the U.S.A. with benefits for environmental, recreation, and many more esoteric effects especially noted. Reference 1 introduces some of the terminology of cost allocation and illustrates the general procedure:

Specific Costs The readily determinable costs of facilities that are clearly for one purpose only should be allocated specifically to that purpose.

Separable Costs These include specific costs plus that part of joint costs traceable solely and clearly to the inclusion of a single purpose in the multipurpose project. This cost for each purpose is defined as the difference between the cost of the multipurpose project and the cost of the same project without the purpose.

Steps in using SCRB (Scrub) method, refer to Table 5-19.

1. The benefits of each purpose are estimated.

2. The alternative costs of single purpose projects to obtain the same benefits are estimated.

3. The separable cost of each purpose is estimated.

4. The separable cost of each purpose is subtracted from the lesser of each purpose's benefits or alternative cost. The lesser figure is used because alternative cost is used in this method only if it represents a justifiable expenditure, that is, if it does not exceed the benefits.

5. From the total cost of the project, all separable costs are subtracted to determine residual costs.

6. Residual costs, designated as joint costs in this method, are distributed in direct proportion to the remainders found in step 4.

7. To determine the cost allocated to each purpose, the separable and distributed costs for each purpose are added. In the case of power, from that sum is subtracted the amount of taxes foregone which was used in computing power costs under steps 2. and 3.

TABLE 5-19

SIMPLIFIED ILLUSTRATION OF APPLICATION OF SEPARABLE COSTS
REMAINING BENEFITS METHOD

(Values in Thousands of Dollars)

ITEM	Project Purposes			
	Flood Control	Irri-gation	Power	Total
1. Benefits	8,000	23,000	35,000	66,000
2. Alternative Single-Purpose Cost	10,000	27,000	30,000	--
a. Lesser of Item 1 or Item 2	8,000	23,000	30,000	--
3. Separable Costs	0	8,000	29,000	37,000
4. Remaining Benefits (Item 2a - Item 3)	8,000	15,000	1,000	24,000
Percentage of Total	33	62	5	100
5. Unallocated Joint Cost (Project Cost - Item 3)	--	--	--	19,000
6. Allocated Joint Cost (percentage x 19,000	6,300	11,800	900	19,000
7. Total Allocation (Item 6 + Item 3)	6,300	19,800	29,900	56,000

Those involved in cost allocation are advised to use the simplest possible method appropriate to the circumstances. The extra cost estimates required for a precise application of the SCRB method are often costly in time and effort to perform. Experience indicates that good cost estimates for alternative single-purpose schemes and for the several different multipurpose types needed for the SCRB method have not always been made.

87

References

1. "Cost Allocation for Multipurpose Water Projects," N.B. Bennett, Jr., Transaction of the American Society of Civil Engineers, Volume 123, 1958, pp 85-100.

2. "Economic Analysis of Agricultural Projects," J.P. Gittinger, a World Bank Publication, Johns Hopkins University Press, Baltimore, Maryland, 1972, Fifth Printing, 1976.

3. "Economic Analysis of Projects," L. Squire and H.G. van der Tak, World Bank Research Publication, Johns Hopkins University Press, Baltimore, Maryland, 1975.

4. "Economic Practices Manual Draft," California Department of Water Resources, Sacramento, California, 1977.

5. 'Guide to the Economic Evaluation of Irrigation Projects," H. Fergmann and J.M. Boussard, Organization for Economic Co-Operation and Development (OECD), 2 rue Andre-Pascal, 75775 Paris Cedex 16, 1976.

6. "Principals and Standards for Planning Water and Related Land Resources," U.S. Water Resources Council, Washington, D.C., April, 1980.

7. "Proposed Practices for Economic Analysis of River Basin Projects," Report to Federal InterAgency River Basin Committee, Washington, D.C., 1950.

8. 'The Capital Budgeting Decision," H. Bierman, Jr. and S. Smidt, MacMillan Publishing Co., Inc., New York, New York, Fourth Edition, 1975.

9. "The Economic Choice Between Hydroelectric and Thermal Power Development," H.G. van der Tak, World Bank Staff Occasional Papers Number One, Johns Hopkins University Press, Baltimore, Maryland, 1966, Fourth Printing, 1974.

CHAPTER 6. PLANNING FOR AND EVALUATION OF ENVIRONMENTAL EFFECTS

Introduction

Once a majority of the population fulfills its basic needs of food and shelter, they turn their attention to satisfying higher level needs related to personal and esthetic desires. Therefore, in the more developed countries, more emphasis is placed on conserving and preserving environmentally significant resources. Although less developed countries are more interested in satisfying basic human needs, such as providing food and shelter for their expanding populations, they are also becoming more concerned about their country's environmental attributes. Therefore, as in the social area, arguments have been put forth that water resource planners should, as a minimum, evaluate an irrigation and drainage project's environmental affects and in some cases formulate plans that specifically meet environmental problems and needs.

If within the planning setting it is determined that environmental concerns should be incorporated into the planning process, it will be necessary to determine to what degree they are to be treated as part of the process. This could range from developing alternatives that meet specific environmental problems and needs, to merely evaluating the environmental affects of each alternative plan.

Environmental Problems and Needs

If it is decided that projects should be formulated to resolve or meet environmental problems and needs it will be necessary to identify and quantify these problems and needs. Following is a list of example problems and needs.

1.   Acquiring and providing for maintenance of land and water in congested areas as open space and greenbelts for parks, for recreational development on flood plains, and for generally preserving riverine areas in cities and towns.

2.   Preserving and maintaining the intrinsic values of flowing streams (wild, scenic, and recreation rivers) for public use, for fish and wildlife habitat, and for visual resource management.

3.   Protecting natural lake environments and providing and managing reservoirs to provide high biological productivity of desirable species.

4.   Protecting beaches and shores of high value and providing for acquisition and management of additional areas where appropriate.

5. Conserving, preserving, or enhancing wilderness, primitive, and natural areas.

6. Preserving and maintaining estuarine and wetland areas that are important feeding and nursery habitat for a variety of plants and animals.

7. Preserving, for future use and enjoyment, areas of natural beauty including, but not limited to, waterfalls, scenic canyons, and outstanding views.

8. Providing for the preservation and interpretation of cultural resources, including archeological, paleontological, and historic aspects that may be affected by the use of water and related land resources.

9. Maintaining or improving the habitat of plants and animals listed or proposed to be listed as threatened or endangered species.

10. Providing for the maintenance of adequate habitat to support diverse populations of plants and animals, including those that contribute to the observation, study, and harvest of fish and wildlife resources.

11. Preserving and maintaining areas that display and illustrate geological processes.

12. Providing for maintenance of certain discrete ecosystems so that they remain in good condition.

13. Providing for improving water quality now below acceptable standards. Providing for maintenance of water quality at a level which provides for present and future uses.

14. Providing for maintenance of air quality at a level which provides for future uses.

15. Conserving productive soil for future use in producing food and fiber, as well as a part of the habitat of man and other species. Providing for the use of land, including prime and unique farmland, within its carrying capacity, instituting wise land use practices so that the land is not degraded, and instituting measures to correct past misuses.

16. Reducing noise so that it is not a hazard to health or does not become a nuisance or problem.

17. Maintaining the aesthetics of project areas and designing future projects and programs so that visual quality is improved rather than impaired.

Once the environmental problems and needs are identified and quantified, they should then be combined with the identified economic and possible social problems and needs. Then plans can be formulated to resolve in varying degrees those problems and needs.

The key to formulating plans that resolve or meet environmental problems and needs depends upon identifying and quantifying environmental problems and needs in the context of the overall planning setting.

While it is often not difficult to identify an environmental problem or need, it is much more difficult to quantify them. For instance, the need for a wild and scenic river may be identified. But how do you quantify this need and how do you evaluate it in relationship to other problems and needs? This is where the statement "in the context of the overall planning setting" must be taken into consideration. It is the planner's responsibility to evaluate or weigh all the problems and needs in the overall context of the planning setting. What this implies is that in some countries where meeting food and fiber needs are more important than meeting environmental needs, more weight will or should be given to meeting food and fiber needs. On the other hand, if the environment is coequal with national economic concerns it should be treated as an equal in the planning process.

Evaluation of Environmental Effects

While identifying and formulating plans to meet environmental problems and needs is dependent upon the planning setting, it is generally good practice from a plan formulation standpoint, to evaluate the environmental effects of all potentially implementable alternatives. In fact, throughout the plan formulation process the evaluation of environmental effects can be used as a basis to reject or retain alternatives for further consideration and analysis.

Environmental Components and Elements

To evaluate environmental effects it is convenient to think in terms of major environmental components and related elements.

As an example, listed below are the components and elements identified by the U.S. Water Resources Council. Although this list represents the U.S. situation, it basically encompasses the major environmental categories that would probably be encountered throughout the world.

A. Ecological Component.

This component includes elements of environmental quality important in water and related land resources planning because of their ecological value. Ecological values are usually defined in national legislation but can be established by decisionmakers at the beginning or as a part of the study.

(1) Biological Resources.

This element includes effects on individuals, species, and populations of living organisms. Both flora and fauna are to be evaluated under this category.

Flora includes plants as individual species, as stands of individual species, and as communities of associated species.

91

Fauna includes animals and their habitats. Species to be evaluated usually include at least those species covered by National legislation. Each important animal group should be considered and its characteristics described under each condition of analysis, i.e., existing conditions, without the project and with the various project alternatives. Species having generally similar life cycles, or habitat requirements, or otherwise logically fitting together for purposes of analysis should be considered together in the analysis.

(2) Ecological Systems.

This element includes the identifiable communities of organisms and their interrelationships with their physiochemical surroundings. Each natural area, such as a watershed, a vegetation and soil type, a tidal salt marsh, a swamp, a lake, or a stream complex, represents an eco-system – an interdependent physical and biotic environment that functions as a continuing dynamic unit. When such natural areas may be lost or otherwise diminished in size or quality, there are corresponding adverse environmental effects borne by society.

(3) Estuarine and Wetland Areas.

An estuary is defined as a semi-enclosed coastal water body having free connection with the open sea within which seawater is measurably diluted with freshwater drained from the land. The estuarine system would include the water, submerged lands, marshes, intertidal lands, and shoreward lands, plus the fauna and flora which are characteristic of such a system.

Wetlands are defined as lowland areas that are permanently or intermittently covered with shallow water, often referred to as marshes, swamps, sloughs, or potholes, generally with emergent vegetation as a conspicuous feature.

(4) Wilderness, Primitive, and Natural Areas.

Wilderness and primitive areas are those areas defined as lands included within or having the potential for inclusion within a National wilderness preservation system, or having similar qualities and characteristics. Such areas should be undeveloped land retaining primeval character and influence, without permanent improvements or human habitation, which is protected and managed (or has the potential for so being) so as to preserve its natural conditions. The area should be of sufficient size so that preservation is practical and using it would not impair it. The area may contain ecological, geological, or other features of scientific, educational, scenic, or historical value. Prairie grasslands and desert areas, as well as forested mountain areas, could be included in this category.

Natural areas are areas that contain rare or unique biotic, geologic, pedologic, or aquatic characteristics, forms, and processes. They may range in size from less than an acre to many thousands of acres. Areas may be set aside for their visual quality or as research natural areas for scientific and educational purposes.

92

B. Physical Component

This component covers the environmental quality aspects of water, air,
land, and sound; the visual quality of landscape resources; and geological
phenomena.

Included under this component are the following elements.

(1) Water Quality

This element includes the chemical, physical, and biological properties
of fresh, brackish, and saltwater measured in terms of its suitability
for a particular use.

(2) Air Quality

This element includes the chemical and physical aspects of air.

(3) Land Quality

This element includes chemical, physical, and biological aspects of
land in relationship to the suitability of the land for particular uses
such as grazing, homesites, cultivation, etc.

(4) Sound Quality

This element covers the effects of sound as it relates to the quality
of the environment.

(5) Visual Quality

This element assesses the benefits from visually attractive landscapes
as well as the adverse effects of features that destroy, disrupt, or
intrude on pleasing settings.

(6) Geological Resources

This element covers areas of geological importance as future mineral
supplies as well as those areas of geological interest that are used in
studying or displaying the development of the earth.

C. Cultural Component

This component covers the human-related structures, places, and objects
of significance in history, architecture, archeology, culture, or
science. The following elements are included in this component.

(1) Archeological Resources

Archeological resources include those material remains such as occupation
sites, work areas, evidence of farming or hunting and gathering, burial
sites, artifacts, and structures of all types of past human life and
activities during prehistoric periods (or during historic periods for
which only vestiges remain).

(2) Historical Resources

Historical resources include those remaining evidences of the origins,
evolution, and development of the Nation, state, or locality. It also
encompasses recognition of places where significant historic or unusual
events occurred even though no evidence of the event remains, or places
associated with a personality important in history.

J. Recreational Component

This component covers water, land, and water-related or land-related
resource areas available for recreation and enjoyment, including areas
identified by National legislation or administrative action for protection primarily for their recreational value.

(1) Streams and Stream Systems

This element refers to any natural water course, whether flowing year-
round or intermittently including reaches of stream between water
development projects.

(2) Beaches and Shores

This element includes land areas adjacent to saltwater, estuarine
areas, freshwater lakes, reservoirs, and streams that provide access to
and from the water. Consideration should be given to access and use
for swimming, boating, fishing, and waterfowl hunting, etc.

(3) Lakes and Reservoirs

This element includes natural and manmade lakes, reservoirs, and other
areas of standing water, except those areas classed as wetlands or
estuaries. If the quantity of water impounded behind a dam or other
structure is materially increased, that aspect should be included in
this analysis.

(4) Open Space and Greenbelts

This element includes natural landscapes used to maximize natural and
spatial values. It should be considered in the same framework as any
other land use which performs economically and socially desirable
functions. These functions are basically: resource production;
preservation of natural and human resources; health, welfare, and well-
being; public safety; transportation corridors; urban and rural development;
and recreation opportunity.

Measurement Criteria

To evaluate an alternative's beneficial and adverse environmental
effects the above components and related elements should be measured
from the standpoint of quantity, quality, and irreversibility considerations. These criteria are defined as follows.

94

A. Quantity

To the extent practicable, the specific environmental resources and aspects evaluated within each element are to be measured and displayed in terms of accepted linear and volumetric units, and/or numbers of animals or places. After such measurements have been made, they may be summarized. It is not necessary to convert the data into monetary units.

B. Quality

To properly indicate public values, a subjective judgment of the quality of the resources covered by each category will generally be required. In cases where quality standards have been promulgated, as for air and water, this information should be included. This evaluation also should consider such factors as the degree to which: people use or would use the resource identified; its availability for continued use; if it might be degraded by use; and its contribution to education, scientific knowledge, and human enjoyment. The quality of the resources considered should be subjectively described by comparing known or projected conditions with conditions that occur at other locations. A valid and reliable scaling technique should be used to measure relative differences in quality among the alternatives considered and future without the project conditions. Reliability and validity of measurements must be shown for each scaling application either by adequate analysis or by reference to related analyses of like applications. Alternative evaluation techniques are discussed later in this chapter.

C. Irreversibility

In evaluating irreversibility, the significance of each anticipated change affecting an environmental resource is determined by considering factors related to the abundance and distribution of each resource, the interrelationships of supply and demand, relative reversibility when compared with other resources, and mitigatory actions.

For instance, some resources can be moved successfully. In some cases, historic structures and artifacts may be relocated. If the resource can be moved, then the significance of the adverse impact is much less than a total loss.

Evaluation Systems

A major issue or problem associated with the evaluation of beneficial and adverse environmental effects is that these effects are of a non-monetary nature. Because of this attribute much thought and research has gone into the development of environmental evaluation systems. However, at this time no universally accepted system has been developed.

Several systems to evaluate the quality parameter have been developed over the past few years. Most of these evaluation systems utilize scaling techniques to evaluate environmental effects.

Following are descriptions of several evaluation systems:

USGS Matrix Method (Leopold et al, 1971).

A matrix (checkerboard) constructed by listing "proposed actions which
may cause environmental impact" along one axis and listing "existing
characteristics and conditions of the environment" along the other
axis; provides for a total of 8,800 possible interactions whose "magnitude' and "importance" are scored on a scale of 1 to 10; characteristics
of the natural environment and cultural/social/human life quality are
considered.

Optimum Pathway Matrix Method (Zieman et al, 1971).

A computerized method of determining an impact index for alternatives
for highway location by evaluating 56 parameters representing economics/
engineering, natural environment, and social/human life quality categories;
parameters are evaluated by multiplying their component value by a
"weighting factor" and "scaling factor" and by using a statistical
error estimating procedure to account for imprecision.

Battelle Memorial Institute Method (Whitman et al, 1971).

A method of comparing projections of the with and without project
conditions by proportioning 1,000 "environmental quality units" among
66 parameters representing the categories of ecology, environmental
pollution, aesthetics, and human interest; the importance of parameters
is determined by the number of "EQ units" assigned; the "EQ unit" score
that an alternative receives is determined from "value functions" which
relate each parameter's total assigned "EQ units" to real-world units
of measurement such as acres, ppm, miles, etc.

Western U.S. Water Plan Method (U.S. Bureau of Reclamation, 1971).

An evaluation system consisting of 18 categories representing considerations of the natural environment and cultural/social/human life quality;
the importance of categories is rated on a scale from 1 to 10; the
magnitude of quantifiable parameters is expressed in acres, miles,
volumes, etc., and the magnitude of unquantifiable parameters is rated
on a scale from 1 to 10.

Northwestern University Method (Gemmel et al, 1972).

A relatively complex system in which a large number of parameters are
evaluated in a series of matrix and vector analyses; the "vector of
impact values" for an alternative is equal to the product of a "matrix
relating system attributes to changes in environmental conditions"
times a "matrix relating changes in environmental conditions to changes
in fundamental human activities" times a "vector describing relative
values of human activities"; sensitivity is measured on a scale from -5
to +5.

Matrix Analysis of Alternatives for Water Resources Development
(Draft Technical Paper, U.S. Army Engineers District Tulsa, CE 1972).

This method can be classified as a matrix method which quantifies all
the impacts and weighs each impact and each objective to give an
overall number as a measure of effectiveness. With this method the
economic, environmental and social well-being accounts are reduced to a
non-dimensional rating system so that the individual economic, environmental and social well-being scores can be added together to determine
an alternative's overall effectiveness.

Battelle Columbus Laboratories (Duke et al 1977).

This method was developed to assist the U.S. Bureau of Reclamation in
implementing the Water Resources Council Principles and Standards.
Four components, 15 categories and 75 evaluation factors are used to
organize environmental measurements into a standard procedure for
displaying the beneficial and adverse effects of water resource development projects on environmental quality. Examples of appropriate
measurements for each evaluation category are provided along with
suggested tables that might be used to display and summarize data used
in the evaluation.

Analysis and Displays

The above evaluation systems provide the framework for evaluating
environmental effects.

While environmental effects are sometimes difficult to categorize as
being beneficial or adverse, it is important that the results of the
evaluation be quantified and displayed to facilitate consideration by
the decision makers.

In evaluating environmental effects, care should be taken to not
evaluate the same values more than once. For instance, the same
resource, such as land area, may be considered in several contexts.
Several measurements may be necessary to assure consideration and
coverage of the various ways in which environmental quality depends on
a single land or water area. Evaluation results from different elements
should not be added together.

A key element in the environmental analyses, as it is with the economic,
and social effects analyses, is the development of a future without the
project "scenario." While this is a difficult task in the economic
area, it is a most difficult task in the environmental area.

Since all alternatives are measured against the "future without" scenario
it is important to develop a future without plan that represents future
improvement or degradation in all the environmental elements.

The "future without" should be included in any display of environmental
effects. Examples of environmental displays are enclosed. Tables 6-1,
6-2, and 6-3 are examples of environmental displays from the Battelle
Columbus Laboratories Report (1977) prepared for the U.S. Bureau of
Reclamation.

TABLE 6-1

EXAMPLE MEASUREMENTS FOR THE BIOLOGICAL RESOURCES CATEGORY

Factor - Measurements	Present Condi- tions	Future Conditions		
		With- out	Alterna- tive A	Alterna- tive B
Aquatic Flora				
Area (acres)	8	9	6	6
Grasses & Shrubs				
Area (acres)	8576	8461	5763	5623
Carrying Capacity in A.U.	1020	1000	690	670
Aquatic Animal Species				
Harvest				
Large-mouth bass (lb/ac)	15	18	75	80
Channel catfish (lb/mile)	10	10	13	13
Terrestrial Animal Species				
Game Harvest				
Mallard (No./mi^2)	7	7	10	11
Pheasant (No./mi^2)	12	10	14	14
Fur Harvest				
Muskrat (pelt/mi^2)	350	350	200	200
Endangered and Threatened Species				
Number of Species				
Plants	2	3	3	3
Animals	1	1	1	1
Areas Used (acres)	43	46	46	46
Unique Biota				
Area (acres)	2	2	2	1
Education and Scientific Values				
Population Groups Expected to				
Change (number)	0	0	4	4
Study Area (acres)	0	0	2000	300
Legal & Admin. Protection				
Area Regulated (acres)	4000	5000	6000	6000

TABLE 6-2

<u>EXAMPLE MEASUREMENTS FOR THE ECOLOGICAL SYSTEMS CATEGORY</u>

Factor - Measurements	Present Conditions	Future Conditions		
		Without	Alter. A	Alter. B
Ecosystem Types				
Grassland (acres	10,069	10,044	7,085.5	6,896
Stream (miles)	7	7	2.5	2
Lake (acres)	57	71	3,057	3,257
Ecosystem Quality				
Climax				
Grassland (acres)	73	70	70.5	67
Wetlands (acres	73	66	60	55
Stream (miles	2	2	0	0
Subclimax				
Grassland (acres)	51	55	50	48
Wetlands (acres)	44	62	45	28
Stream (miles)	5	5	2.5	2
Lake (acres)	18	18	3,018	3,257
Disclimax				
Grassland (acres)	9,945	9,419	6,965	6,781
Lake (acres)	39	53	39	39
Uniqueness				
Area (acres)	23	23	23	23
Educ. & Scientific Value				
Area (acres)	120	120	120	120
Legal & Admin. Protection				
Area Protected (acres)	50	80	4,500	5,000

99

TABLE 6-3

EXAMPLE MEASUREMENTS FOR THE ESTUARINE AND
WETLAND AREAS CATEGORY

Factor - Measurements	Present Conditions	Future Conditions		
		Without	Alter. A	Alter. B
Estuarine Areas	0	0	0	0
Wetland Types (acres)				
Inland Fresh Meadows	12	12	8	8
Inland Shallow Fresh Marshes	15	26	12	6
Inland Deep Marshes	20	20	15	11
Inland Open Fresh Water	70	70	70	70
Productivity				
Waterfowl Use (No. broods/mi^2)	11	10	4	3
Educ. & Scientific Value				
Area (acres)	22	22	16	16
Legal and Admin. Protection				
Area Protected (acres)	30	70	105	95

CHAPTER 7. SOCIAL IMPLICATIONS AND ANALYSIS

Introduction

Although they are not often quantified, social factors play an important
role in water resource development, and in particular in irrigation and
drainage developments throughout the world. In fact, for some time and
with increasing intensity, arguments have been advanced that transi-
tional economic analysis which emphasized economic efficiency should be
modified to explicitly reflect social preferences. That is, irrigation
and drainage projects should be formulated to meet social objectives
such as raising the income of a particular segment of the population, or
a particular region of a country. This argument arises from the need to
satisfy basic human and social needs of food, fiber, shelter, security,
belonging, self-esteem, accomplishment, etc.

Abraham Maslow, reference 1, describes these needs and divides them into
five distinct categories. These categories are as follows:

o   Psychological - this category represents the need for food,
     water, clothing, shelter, sleep, air, reproduction, etc.
o   Security - the need for protection from physical harm, insurance
     of continuing income, and employment.
o   Social - a need for a sense of belonging to a group and
     acceptance by other people.
o   Ego - need for the achievement of independence, self-esteem,
     the deserved respect of peers, recognition.
o   Self-realization - which represents the need for a sense of
     accomplishment and achievement of full capability, acceptance
     of new challenges, and broadening of horizons of interest.

Maslow's theory goes on to state that as lower order needs are satisfied,
the next need becomes the objective. Therefore, countries where most of
the population is sustaining life on a minimal diet, or experiencing
water-related diseases due to poor water quality or treatment facilities,
will support projects that meet the basic psychological needs of food,
water, and sanitation, etc.; and it is with the psychological and
security needs that economic variables play an important role.

As the psychological and security needs are satisfied, the population
begins to turn its attention toward higher order needs, that is ful-
filling an individual's social, ego, and self-actualization needs. It
is in these stages that the individual becomes concerned with the
satisfaction of personal, cultural, esthetic, and environmental desires.
At this stage of an individual's life, projects which meet these
objectives will be preferred over those that might provide food or fiber
for export, hydroelectric power, etc. Generally, these higher order
needs are more evident in developed countries. Therefore, it is
very important from the planner's viewpoint to determine the personal

needs or expectations of the majority of the population, and in some cases, the special interest groups at the beginning of a study, and to factor these needs into the planning process.

Evaluation of Social Effects

Before discussing the evaluation of social effects, the difference between planning to meet social objectives and the evaluation of social effects must be differentiated. As was indicated earlier, considerable discussion has surrounded the need to develop irrigation and drainage projects that resolve social problems or meet social needs. Although this has not been explicitly treated in the past, it has been implicitly involved to some extent in most irrigation and drainage projects that have been implemented.

In general, most irrigation and drainage projects in the past have been developed to improve or stabilize a local or regional economy as well as to improve a nation's ability to meet its own basic human needs of food, fiber, etc. Therefore, the planner must determine at the beginning of the study if certain specific social objectives are to be met.

As with the National, regional economic environmental sectors, if projects are to be formulated to meet social objectives, then the social problems and needs must be identified and quantified. These problems and needs could take the form of formulating plans that would benefit minority social groups, relieve adverse social effects in chronically depressed areas, stabilize regional social systems or economics, etc. On the other hand, the evaluation of each alternative's monetary and nonmonetary social effects will provide the decision maker with information that can be utilized to determine the relative merits of each alternative from a social effects standpoint. In fact, it is suggested that the planner utilize the social effects information throughout the planning process to reject or retain alternatives for further study.

An irrigation and drainage project's social effects are measured in both monetary and nonmonetary terms. The monetary effects are often referred to as secondary or indirect benefits to the regional or local economy. The measurement of these indirect or regional benefits are discussed elsewhere in the report. The remainder of this chapter focuses on the evaluation of the nonmonetary social effects.

Components to be Evaluated

The following nonmonetary social effects components should be evaluated.

    o   Individual Effects
    o   Community Effects
    o   Area Effects
    o   National Emergency Preparedness Effects
    o   Aggregate Social Effects

The categories selected and depth of the evaluation will be project dependent and it is the responsibility of the planner to evaluate this requirement early in the planning process. The components and elements associated with each component are described below:

Individual Effects

This component is concerned with the impacts of alternative plans as they are experienced at the most basic level of society, that of the individual and family. The component focuses on the effect of each alternative plan on the personal health and well-being of members of communities in the planning area. It also assesses whether the plan's direct effects will contribute to or detract from the quality and stability of family life. Major evaluation elements of this component include the following:

Life, protection, and safety deal with the effects of natural disasters on life and property and the possible diminution of such effects by alternative plans.

Health deals with both the health of individuals in the communities of the planning area, and the services and facilities available to deal with health problems.

Family and individual deal with the potential direct and indirect impacts of alternative plans on the well-being and personal satisfaction of individuals in the community, and on the structure and stability of family life.

Attitudes, beliefs, and values deal with the attitudes of individuals in the communities towards alternative plans, and the extent to which these plans affect people's attitudes towards other concerns, including themselves and their lifestyles, their community, the environment, Government agencies, etc.

Environmental contributions deal with the community member's interaction with their environment and the extent to which alternative plans may affect their interaction, i.e., by contributing to or detracting from individuals' enjoyment of their environment.

Community Effects

This component deals with the impacts which might occur at a higher level of aggregation than those in the first component, but which are still limited to the confines of the community or communities which would be affected by the alternative plans. Here the focus is on groups of people, their activities as members of different groups, the various aspects of the community which affect their lives, and the informal and formal institutions which serve them. Equally important is the nature of the community itself as an entity, e.g., its composition, its internal structure, its processes and functioning. If one of the potential impacts of the plan is to bring a large influx of population into the area, the resulting changes in the community in terms of size, ethnicity, age patterns, etc., are important. These are likely to create further changes in social components at the individual and area levels, as well as at the community level. Major evaluation elements of this component include the following:

Demographic deals with the structure of the community in terms of size, ethnicity, age, and sex distribution, etc.

Education deals with educational institutions within the community. These include primary, secondary, and postsecondary academic education, as well as vocational training, and their activities and capacities.

Government operations and services deal with the structure, size, and complexity of the formal institutions of local government, and the services which government provides to members of the community.

Housing and neighborhood deal with the quality and quantity of housing in the community, as well as the condition of its neighborhoods.

Law and justice deal with the criminal justice system in the community, and the possible impacts in terms of criminal and civil violations which might result, either directly or indirectly, from the implementation of a water resource plan.

Social services deals with the public and private sector services available to various population groups (e.g., childern, youths, elderly persons) within the community.

Religion deals with the religious structure of the community, and the extent to which religion affects the lives of community members.

Culture deals with the ethnic composition of the population and its associated values and folkways, as well as archeological sites, artifacts, and other cultural materials found in or near the community.

Recreation deals with recreational facilities, and land and water areas in or near the community used for public recreation, and to changes in recreational uses which might occur as a result of the implementation of a water-related plan.

Informal organizations and groups deal with community groups which are not a part of the local governmental or institutional structure, including fraternal organizations, advocacy groups, religious or ethnic societies, environmental groups, neighborhood groups, or other informal associations which might be affected by the implementation of a water resource plan.

Community and institutional viability deal with the capacity of a community's institutions to meet demands for a range of services, on the interrelationships among community institutions, and on the members' views of the community.

Area Effects

Aspects of this component may be applicable at the community level, but the categories grouped under it may involve more than one community, or be important enough to pervade several aspects of community life at once. Consideration of transportation, for example, cannot be limited to the community level alone, since transportation by definition

104

frequently relates to exchanges among communities. Likewise, a discussion
of the economic base of a community and the changes which might occur
in it as a result of a water resource plan's implementation, must give
consideration to the more far-reaching effects which accompany changes
in the economy, such as changes in housing, public and private sector
services, etc. The categories discussed here are thus of a fairly
broad, comprehensive nature, although their focus will sometimes be
within the community context. As data and analysis developed for this
component are closely related to those required for the RD Account,
care should be taken to insure their close coordination. Major evaluation of this component include the following:

Employment and real income deal with the means by which
people in the area earn their living, and the amount of income which
they receive.

Welfare and financial compensation deal with the quality and
quantity of benefits, including money and service provided to people
who, for various reasons, are unable to support themselves.

Communications deal with various inter- and intracommunity
methods of communications, including personal and media methods. It
considers how a water-related plan may affect communications and may be
affected by them.

Transportation deals with public and private transportation,
both within and between communities, as well as with the condition and
location of roads and waterways used for various types of transport.

Economic base deals with those activities which provide the
basic employment and income on which the rest of the regional economy
depends. It considers the changes which might occur in basic industry
and agriculture sectors as a result of water-related projects, and the
social impacts of these changes on the community.

Planning deals with the period during which alternative plans
are formulated and analyzed, and the effects which the planning process
might have on the communities which would be impacted by plan implementation.

Construction deals with with the potential short- and long-
term effects of construction, ranging from temporary noise and worker
influx to long-range economic and social impacts.

National Emergency Preparedness Effects

An important consideration is the way in which the changes brought
about by a plan would have on selected national impacts; that is, the
impacts and effects on the region's contribution of national preparedness
to deal with emergencies. In many cases, these changes will, in fact,
be limited to the geographical area covered by the plan. In others,
however, changes could contribute to or detract from national viability,
especially in the case of emergencies such as drought, floods, or
attacks by hostile forces. For example, plans which will contribute to

105

the stability of a water supply over a large area, or which will enable the conservation of scarce fuels have important implications for national goals. Major evaluation elements of the component include the following:

Water supplies deal with questions of the quantity, quality, stability, distribution, etc., of water.

Food production deals with provision of reserve food production potential, as well as questions of exportability of crops.

Power supplies deal with questions concerning the quantity, stability, and responsiveness of available power.

Water transportation deals with tonnage support capacity and network location characteristics, as well as the service area covered by water transportation routes.

Scarce fuels deals with the use of abundant fuels to conserve scarce ones, and the use of alternative power supplies.

Population dispersion deals with potential for resettlement of areas in the event of a national emergency.

Industrial dispersion deals with the potential for dispersion and relocation of industry in the event of a national emergency.

Military preparedness deals with the potential of a given area to house military bases and support military operations in a national emergency.

International treaty obligations deal with the question of boundaries, use of common bodies of water, treaty obligations governing quantity or quality of water crossing international boundaries, etc., among bordering countries.

Aggregate Social Effects

This component involves the aggregation of potential impacts of water resource projects as they relate to social effects as a whole. It considers the more general, overall social implications of alternative water resource plans which may not have been fully expressed in evaluations of the other components. The measures of impact will, in this case, be of an almost entirely qualitative nature.

The ratings given to potential effects will be based, in large part, on professional judgment and interpretations resulting from observations made at the more operational levels of data and evaluation for the preceding four components. Major evaluation elements of the component include the following:

Quality of life deals with the physical and psychological well-being of the individual and family, and with their perceptions of the opportunities for further development of individual and family life in the future.

106

Relative social position concerns the extent to which the various social benefits and adverse effects of plan implementation would be equitably distributed among various individuals or groups in the community, and the capacity of individuals and groups to bear social costs.

Social well-being deals with the overall impacts on the character and capacities of the community and its institutions, both formal and informal.

Evaluation Process

In terms of a process, the following steps should be used as a guide in evaluating social effects. Five general steps are involved in conducting a social assessment.

Step 1.   Summarize the water development history of the planning area and what an alternative plan entails.

Step 2.   Summarize the history and present day social characteristics of a planning area. This step provides a narrative describing the history and present status of the human environment.

Steps 1 and 2 should be prepared in close coordination with the rest of the planning team to avoid duplication.

Step 3.   Forecast the future conditions without a plan, the impacts of each alternative plan, and then assess the beneficial and adverse social effects for each alternative plan.

Step 4.   Perform tradeoff analysis by comparing the social effects of the alternate plans with one another.

Step 5.   Develop a set of conclusions regarding the best course of action from the social perspective for input into the other portions of the planning process.

Data

The value of the results of the social effects evaluation are directly related to the quantity and quality of the data used. Therefore, in establishing a social effects evaluation program the following factors should be considered:

o   The level of the planning study (i.e., appraisal, feasibility, advance planning, special, etc.).

o   The amount, quality, and currentness of social data available from previous studies.

o   The primary purpose of the social assessment, that is, is the assessment being prepared to support other documents, etc.

o   The geographic size of the study area; the number of people potentially affected, and anticipated degree of impact; the number of different functions involved, that is irrigation, power, municipal and industrial, flood control, etc.; the size, relative location of the communities potentially affected; the presence of native or ethnic people, or other groups of special concern in the study area, etc.

Data utilized in the social effects evaluation process can be separated into two categories: Primary and Secondary. The need for Primary or Secondary data will be dependent upon the level of study and the degree of uncertainty or importance surrounding the social issues of the study. Secondary data is defined as data coming from other sources and usually refers to census data. In some cases it may also refer to special studies and reports prepared by or for agencies having responsibility for a particular region or urban area in a country. Primary data refers to data collected specifically for a project area, and involves the following efforts: development of questionnaires and/or interview programs. Social and behavorial scientists should be utilized to identify, develop techniques, collect and analyze this type of data.

Methods

Evaluation of a project's social affects is a formidable task. The first and major issue to confront the planner is that the information to be evaluated is nonmonetary. Although hard data, that is, facts and figures, are available, they cannot, without great difficulty, if at all, be translated into monetary values. Furthermore, the nonmonetary data can be categorized into objective and subjective data. The objective, or hard data referred to can be subjected to normal statistical analysis and relative comparisons made.

The subjective data leads us into a different set of evaluation procedures involving psychometric and sociometric methods. These are scientifically substantiated methods that can be utilized to collect, analyze and interpret subjective data. A full treatment of these methods in this report is not possible; however, references 2 and 3 treat these techniques in more detail.

Presentation of Results

In presenting the results of a social analysis, two steps are involved. The first step is to present a "future without the project" scenario. The second step involves comparing the social effects of each alternative to the "future without the project" to determine net effects of each alternative.

Development of the "future without" scenario is a difficult but very important task.

It basically involves projecting significant demographic, economic, technological, and environmental conditions into the future. Because this is an important task it is suggested that: the significant variables be tested from a sensitivity standpoint to determine what variables are the most significant, and a range of projections be considered.

108

A key element in the evaluation of social effects is the presentation of the results of the study. Using the components and elements described earlier, Tables 7-1, 7-2 and 7-3 show examples of how these data can be evaluated, arranged and presented. These examples are taken from reference 3.

References

1. "Motivation of Personality," A.M. Maslow, New York, Harper & Row, 1954.

2. "A Guide to the Preparation of the Social Well-Being Account, Social Assessment Manual," S.J. Fitzsimmons, L.I. Stuart, & P.E. Wolff, prepared for the Bureau of Reclamation by Abt Associates Inc., 1975.

3. "Scaling Impacts of Alternative Plans," C.A. Brown, R.J. Quinn, and K.R. Hammond prepared for Resource Analysis Branch, Division of Planning Technical Services, Bureau of Reclamation by Center for Research on Judgment and Policy, Institute of Behavioral Science, University of Colorado, Boulder, CO, June 1980.

TABLE 7-1

BENEFICIAL & ADVERSE EFFECTS ON SOCIAL WELL-BEING UNDER THE PLAN FOR

_____

Items in Evaluation Categories	Measures Of Impact	Effects ++,+,0,-,--
I. INDIVIDUAL, PERSONAL EFFECTS		
A. LIFE, PROTECTION & SAFETY		
Change in State Variables		
1. No. of persons served by each agency, by services received		
2. Property loss due to water-related natural disasters		
3. Life loss due to water-related natural disasters		
Change in Relevant Condition Variables		
4. Climatic trends over next decade (based upon present trends)		
Change in System Variables		
(Quantity & Structure) 5. No. & type of agencies which deal with natural disasters (e.g., Civil Defense, flood control agencies, etc.)		
6. Staff size & budget of agencies which deal with natural disasters		
7. Location of agencies which deal with natural disasters (distance from pop. centers of communities under consideration for plan implementation)		
(Function & Purpose) 8. Governmental level at which agencies dealing with natural disasters are located (e.g., local, county, state)		
9. Staff types of agencies dealing with natural disasters (i.e., professional or volunteer)		
10. Services provided by agencies which deal with natural disasters (e.g., financial, medical, etc.)		
11. Coordination or conflict among agencies dealing with natural disasters		
(Quality & Character) 12. Accessibility of agencies which deal with natural disasters		

TABLE 7-1 CONTINUED

BENEFICIAL & ADVERSE EFFECTS ON SOCIAL WELL-BEING UNDER THE PLAN FOR

_____

Items in Evaluation Categories	Measures Of Impact	Effects ++,+,0,-,--
A. LIFE, PROTECTION AND SAFETY (cont.)		
13. Efficiency and effectiveness of agencies which deal with natural disasters		
Change in Other Variables (Specify)		
B. HEALTH		
Change in State Variables		
1. Morbidity levels, by type of disease		
2. Mortality levels, by type of mortality (type of disease, accident, natural causes, etc.)		
Change in Relevant Condition Variables		
3. Distribution of pollution levels in terms of noxious stimuli (air, noise, water, land)		
4. No. of disease-carrying organisms		
5. Provision of various sanitation services (sewage, garbage disposal, etc.)		
Change in System Variables		
(Quantity & Structure) 6. No. of outpatient care facilities		
7. No. of in-patient beds per 1,000 pop.		

TABLE 7-2

A SUMMARY COMPARISON OF THE BENEFICIAL AND ADVERSE EFFECTS
OF THE ALTERNATIVE PLANS FOR THE SOCIAL WELL-BEING ACCOUNT

Items in Eval. Categories	Present Conditions	Beneficial & Adverse Effects of Plans					
		NO Plan	NED Plan	EQ Plan	ALT "X"	ALT "Y"	RECCO Plan
1. INDIVIDUAL, PERSONAL EFFECTS							
A. LIFE, PROTECTION & SAFETY							
Change in State Variables							
1. No. of persons served by each agency, by services received							
2. Property loss due to water-related natural disasters							
3. Life loss due to water-related natural disasters							
Change in Relevant Condition Variables							
4. Climatic trends over next decade (based upon present trends)							
Change in System Variables							
(Quantity & Structure)							
5. No. & type of agencies which deal with natural disasters (e.g., Civil Defense, flood control agencies, etc.)							
6. Staff size and budget of agencies which deal with natural disasters							
7. Location of agencies which deal with natural disasters (distance from pop. centers of communities under consideration for plan implementation)							
(Function & Purpose)							
8. Governmental level at which agencies dealing with natural disasters are located (e.g., local, county, state)							

TABLE 7-2 CONTINUED

A SUMMARY COMPARISON OF THE BENEFICIAL AND ADVERSE EFFECTS
OF THE ALTERNATIVE PLANS FOR THE SOCIAL WELL-BEING ACCOUNT

Items in Eval. Categories	Present Conditions	Beneficial & Adverse Effects of Plans					
		NO Plan	NED Plan	EQ Plan	ALT "X"	ALT "Y"	RECCO Plan
I. INDIVIDUAL, PERSONAL EFFECTS							
A. LIFE, PROTECTION & SAFETY							
9. Staff types of agencies dealing with natural disasters (i.e., professional or volunteer)							
10. Services provided by agencies which deal with natural disasters (e.g., financial, medical, etc.)							
11. Coordination or conflict among agencies dealing with natural disasters							
(Quality & Character) 12. Accessibility of agencies which deal with natural disasters							
13. Efficiency and effectiveness of agencies which deal with natural disasters							
Change in Other Variables (Specify)							
B. HEALTH							
Change in State Variables							
1. Morbidity levels, by type of disease							
2. Mortality levels, by type of mortality (type of disease, accident, natural causes, etc.)							
Change in Relevant Condition Variables							
3. Distribution of pollution levels in terms of noxious stimuli (air, noise, water, land)							

113

TABLE 7-2 CONTINUED

A SUMMARY COMPARISON OF THE BENEFICIAL AND ADVERSE EFFECTS
OF THE ALTERNATIVE PLANS FOR THE SOCIAL WELL-BEING ACCOUNT

Items in Eval. Categories	Present Conditions	Beneficial & Adverse Effects of Plans					
		NO Plan	NED Plan	EQ Plan	ALT "X"	ALT "Y"	RECCO Plan
I. INDIVIDUAL, PERSONAL EFFECTS							
B. HEALTH							
4. No. of disease-carrying organisms							
5. Provision of various sanitation services (sewage, garbage disposal, etc.)							
Change in System Variables							
(Quantity & Structure) 6. No. of out-patient care facilities							
7. No. of in-patient beds per 1,000 pop.							

*Conditions (Above to below Average: AA,A,BA); Effects (Beneficial to Adverse: ++,+,o,-,--).

114

TABLE 7-3

SUMMARY OF EFFECTS ACROSS PLANS

		NO Plan	NED Plan	EQ Plan	ALT "X"	ALT "Y"	RECCO
Components of the Social Well-Being Account[a]	I. Individual, Personal Effects						
	II. Community, Institutional Effects						
	III. Area, Socio-Economic Effects						
	IV. National, Emergency Preparedness Effects						
	V. Aggregate Social Effects						
Trade-offs[b]	Direct vs. Indirect effects						
	Short-Term vs. Long-Term Effects						
	Geographic Distribution						
	Special Groups Affected						

[a]Rate ++ (very positive); + (positive); o (neutral); - (negative); -- (very negative)

[b]Brief verbal descriptions (e.g., direct positive effects on recreation vs. indirect negative effects on employment).

CHAPTER 8. PLAN SELECTION PROCESS

The culmination of the planning process is the selection of a plan for implementation. At this point in the study, the technical and institutional aspects of the planning process are brought together and a plan is recommended for implementation, or two or three are recommended for consideration by the decision makers.

Technical Aspects

The technical aspects associated with the planning process which should be summarized and displayed for the decision makers include:

o   Physical and programatic features
o   Technical performance
o   National economic and regional economic benefits and costs
o   Environmental and social beneficial and adverse effects

Physical and programatic features includes all the physical structures and programs necessary to implement the plan. This would involve, at a minimum, identifying the specific facilities, training programs, equipment, land acquisition requirements, infrastructure, etc., needed to implement a specific plan.

Technical performance is a measure of an alternative's physical outputs including a comparison of how well each alternative resolves the identified problems and meets projected needs.

National economic and regional economic benefits and costs are a summary of the costs needed to implement all the physical and programatic features of each alternative and the monetary benefits associated with each physical output.

Environmental and social beneficial and adverse effects summarizes each alternative's net contribution or impacts to the environmental and social aspects of the planning setting.

The format and information to be summarized and displayed is dictated by the planning setting and decision makers participating in the selection process. To assist the planner in selecting the material to be summarized the decision makers should be consulted and sensitivity analysis should be utilized to determine the more significant variables.

Institutional Aspects

The institutional aspects of an irrigation and drainage project encompasses the legislative, judicial, standard practices or local customs, and political aspects of the planning setting.

117

In most cases, legislative, judicial, and traditional or local customs
can be incorporated or accounted for in the planning process. That is
international treaties, compacts, local laws, water right laws, local
customs, etc., are incorporated into the technical studies. For
instance, treaties, compacts, water rights, etc., are taken into
consideration in determining the amount of water available for development from both a water quantity and quality standpoint. These constraints, although subject to interpretation, or in some cases, further
adjudication, are relatively easy to incorporate into the technical
aspects of the planning process. Although it is relatively easy to
incorporate these constraints, it is also the responsibility of the
planner to recognize when these constraints are contributing to unrealis-
tic solutions. In this case the planner should develop alternatives
that address these constraints from different perspectives.

Selection Process

Who makes the actual selection is an important aspect of the planning
process and is specifically related to the planning setting and, in
particular, the decision makers associated with the planning setting.

Following is a partial list of potential decision makers that might be
encountered by the planner:

    o   A private citizen or citizens
    o   Municipal officials
    o   County officials
    o   State officials
    o   Regional officials
    o   National officials
    o   Lending institutions
    o   Corporations
    o   Public non-Federal entities
    o   Small business enterprises

and in most cases, combinations of the above.

Generally speaking, the extent to which the planner will carry out the
selection process and the format in which the results of the study are
presented will depend upon who the decision makers are. Therefore, in
some cases a plan may be submitted for approval; and in other cases,
several plans may be submitted for consideration and selection by
designated decision makers.

The planners role in the above process, which is often referred to as
the political aspects of a study, is also dictated by the planning
setting and the related decision makers.

Therefore, the planners participation may range from a presentation of
the study results to a head of a local, state, regional or national
government to the development, implementation, and participation in a
well-structured public involvement program.

118

Based on the above, it can be seen that it is very important for the planner to identify the decision maker or makers early in the planning process. In so doing, the planner should not assume that the project sponsor is necessarily the only decision maker to contend with, but that due to international, national, state, and local laws, treaties, regulations, compacts, customs, etc., there are many other individuals that will need to be consulted as a part of the selection process. The results of these consultations should also be summarized and presented to the decision maker or makers for use in selecting a plan to resolve real problems and projected food and fiber needs.

GLOSSARY

ACCEPTABILITY TEST--The workability of the plan within known institutional constraints and its acceptance by project sponsors, other
interested publics, and by the private and public decision makers
that will ultimately have responsibility for authorizing, funding
and/or implementing the project.

ADVANCE PLANNING--Preconstruction planning before authorization of
a project for construction.

ALKALINIZATION--The process whereby the exchangeable sodium content
of a soil is increased.

ALLOTMENTS (FUNDING)--Funds made available for specific projects or
activities.

ALTERNATIVE COST--The cost of a single purpose alternative means of
providing the same benefits. The alternative may be a single
purpose project at the same or a different site.

AMORTIZATION--To repay a debt by means of a sequence of equal payments.
Part of each payment is used to pay the interest due at the time
it is made and the balance is applied to the reduction of the
principal.

ANNUAL--Occurring once during, or accumulated over, a consecutive
12-month period of time for which the beginning date is identified.

ANNUAL EQUIVALENT COST--The sum of the annual equivalent of the
investment (capital cost), the annual equivalent operation and
maintenance costs, and the annual equivalent of major replacement
costs.

ANNUAL FINANCIAL COST--The sum of the annual equivalent of the fixed
cost, the annual operation and maintenance costs, and the annual
equivalent of major replacement costs.

ARABLE LANDS--Lands capable of being cultivated and suitable for the
production of crops.

The U.S. Bureau of Reclamation is required by law to define arable
land as: "Land which, in adequate units and when properly provided
with the essential improvements of leveling, drainage, buildings,
irrigation facilities and the like, will have a productive capacity,
under sustained irrigation agriculture, sufficient to: meet all
production expenses, including a reasonable return on investment;
repay reasonable irrigation and improvement costs; and provide a
satisfactory level of living for the farm family."

121

ASSOCIATED COSTS--The costs of the goods and services, over and above those included in project costs, needed to make the immediate products or services of the project available for use or sale.

AVERAGE ANNUAL DAMAGES--The weighted average of all damages that would be expected to occur yearly from floods, drought, etc. under specified economic and development conditions. Such damages are computed on the basis of the expectancy in any one year of the amounts of damage that would result from events throughout the full range of evaluation period.

AVERAGE ANNUAL YIELD (WATER)--The average annual supply of water produced by a given stream or water development over a period of time.

AVERAGE CROP YIELD--The amount of production per unit area that is received on the average, when taking into account production hazards (drought, flooding, hail, insects, etc.).

BASE PERIOD--A period of time specified for the selection of data for analysis. The base period should be sufficiently long to contain data representative of the averages and deviations from the averages that must be expected in other periods of similar and greater length. As an example, the U.S. Weather Service computes values of average, heavy, and light monthly precipitation from data observed during the base period 1931-1960.

For groundwater studies, the base period should begin and also end at the conclusion of a dry trend so that the difference between the amount of water in transit in the soil at the end of the base period is minimal.

BASIC DATA--Records of observations and measurements of physical facts, occurrences, and conditions, as they have occurred, excluding any material or information developed by means of computation or estimate. In the strictest sense, basic data include only the recorded notes of observations and measurements, although in general use it is taken to include computations or estimates necessary to present a clear statement of facts, occurrences, and conditions.

BENEFICIAL USE OF WATER--The use of water for any purpose from which benefits are derived, such as domestic, irrigation, or industrial supply, power development, or recreation.

BENEFIT-COST RATIO--The arithmetic proportion of estimated average annual benefits to average annual costs, insofar as the factors can be expressed in monetary terms. It is a measure of the degree of tangible economic justification of a project.

BENEFITS--Increase or gains, net of associated or induced costs, in the value of goods and services which result from conditions with the project, as compared with conditions without the project. Benefits include tangibles and intangibles and may be classed as primary or secondary.

BENEFITS, DRAINAGE--Increase or gain in the value or quantity of goods and services directly attributable to drainage.

BENEFITS, IRRIGATION--Increase or gain in the value or quantity of goods and services directly attributable to irrigation.

BENEFITS, MONETARY--Increase or gain in the value or quantity of (tangible) goods and services directly attributable to the project.

BENEFITS, PRIMARY--Identifiable gains, assets or values directly resulting from any program or project - values of readily measurable products or services - such as increases in net income from changes or more intensive use of property made possible by irrigation, drainage or flood prevention; reduction in damages from erosion, floodwater, and sedimentation; value of downstream benefits that might accrue as a direct result of the works of improvement.

BENEFITS, REDEVELOPMENT (EMPLOYMENT BENEFITS)--Benefits accruing from the use of unemployed or underemployed labor during construction of project measures.

BENEFITS, SECONDARY--Values added over and above the value of immediate products and services that result from subsequent processing such as increases in income by suppliers of goods and services as a result of increased activities made possible by the project; higher employment by related, allied or non-allied businesses; new businesses and industry.

CAPITAL COST--The installation cost plus interest during construction. Also called INVESTMENT. See INSTALLATION COSTS.

CAPITAL EXPENDITURES--Outlays for plant and equipment which are normally charged to fixed asset accounts.

CAPITALIZED COST--The first cost of an asset plus the present value of all renewals expected within the planning horizon.

CAPITAL RECOVERY PERIOD--The period of time required for the net returns from an outlay of capital to equal the investment.

COMPLETENESS TEST--The investments and institutional actions required to implement each alternative in order to assure that the alternative's projected contributions to national, regional, and local objectives can be fully realized.

COMPREHENSIVE PLAN--A plan for water and related land resources development, that consider all economic, environmental, and social factors and provides the greatest overall benefits to the region as a whole.

COMPREHENSIVE BASIN STUDY--A study for the development of the water and related land resources of a basin to make the best use of such resources to meet the basin's needs and make the greatest long-time contribution to the economic growth and social well-being of the people of the basin and the nation.

CONJUNCTIVE USE (WATER)--The integrated use of surface and subsurface water supplies and facilities, normally involving storage of surplus waters when available, for use during periods when water supplies are deficient.

CONSTANT DOLLARS--The real value in dollars, with price of goods and services remaining constant; usually expressed over a period of time from a base year (having value equal to 100). The effect of using constant dollars is to remove changes in value of the dollar due to either inflation or deflation.

CONSUMPTIVE USE--Water consumed by vegetative growth in transpiration and building plant tissue, and water evaporated from adjacent soil, from water surface, and from foilage. It also includes water similarly consumed and evaporated by urban and nonvegetative types of land use.

COST-DEMAND RELATIONSHIP--The relationship between demand for a product and its cost. As it relates to agricultural water, it is the relationship between the cost of water to the farmer and the amount he will demand.

COSTS, PLANNING--All costs incurred in the preparation of an acceptable plan which includes the steps of reconnaissance (field investigation), prefeasibility (preliminary investigation) and feasibility (plan preparation).

COSTS, LAND RIGHTS--Cost of purchase or otherwise obtaining the right to the use of land including legal and administrative costs. Included are the costs of moving or altering transportation routes and utilities.

COSTS, WATER RIGHTS--Costs associated with the acquisition of the right to the use of a quantity of water necessary to construct, operate and maintain the project.

COSTS, CONSTRUCTION--The cost of constructing a project including material, equipment and labor.

COSTS, RELOCATION--The costs that are applicable to displacement of persons, businesses, facilities, farm operations, etc.

COSTS, PROJECT ADMINISTRATION--The cost of personnel services in the administration of contracts. Costs of construction permits and inspection are included.

COST ALLOCATION--The process of distributing project costs equitably among the various purposes served by the project.

COST SHARING--The process of making contributions, by those benefitting from a project or program, towards the cost of that project or program.

COST (OR BENEFIT) STREAMS (FLOWS)--A list of costs (or benefits) in vertical columns opposite the project year in which they occur.

124

CRITICAL CONSUMPTION LEVEL--The level of per capita income in a country
at which the real resource cost incurred by Government by project
investment, and the social benefit enjoyed by the beneficiary as a
result of marginal increase in consumption, are exactly offsetting.
Under this definition, Government (society as a whole) gains
little or nothing by taxing persons with incomes equal to or less
than this income level.

CROP MARKET OUTLOOK--The prospect of demand for crops based on per
capita consumption of food and fiber, areas' share of national
production, and projections of foreign demand.

CROP ROTATION--The practice of growing different crops in succession
on the same land chiefly to preserve the productivity capacity and
fertility of the soil.

CURRENT NORMAL AGRICULTURAL PRODUCTION--A concept used to describe
estimates which conform to a consistent pattern. For example,
estimates of acreage, production, price, and value of crop or
livestock production based on current production technology and
practices, from which abnormalities caused by weather and other
hazards have been removed. It may also be described as a geo-
metrically weighted moving average of production.

CURRENT NORMALIZED PRICES--Weighted averages of agricultural commodity
prices over the last five-year period, with greater weights placed
on the more recent prices.

DEEP PERCOLATION--Water that moves below the root zone so that it will
not be consumptively used by plants or evaporated from the soil.

DEMAND--The numerical expression of the desire for goods and services
associated with an economic standard for their attainment.

CURRENT DEMANDS--The numerical expression of present desire for
goods and services associated with the current economic standard
for their attainment.

PROJECTED DEMANDS--The gross estimated shares of the national
requirements for goods or services that are set as goals. Projected Demands are forecast for future times such as 20 and 50
years hence.

DISCOUNTS (DISCOUNTING)--The process of converting future values,
benefits or costs to present values by the use of a present value
of one discount factor for a specific interest rate.

DRAINAGE--(1) The removal of excess surface or groundwater from land
by means of surface or subsurface drains, (2) The effect of soil
characteristics which regulate the east or rate of natural drainage.

DRAINAGE AREA--The drainage area of a stream at a specified location,
is that area, measured in a horizontal plane, which is enclosed by
a drainage divide.

DRAINAGE BASIN--A part of the surface of the earth that is occupied by a drainage system, which consists of a surface stream or a body of impounded surface water together with all tributary surface streams and bodies of impounded surface water.

DRAINAGE DIVIDE--The rim of a drainage basin.

ECOLOGICAL IMPACT--The total effect of a change, either natural or manmade, in an environment upon the ecology of the area.

ECONOMIC BASE STUDY--A study which evaluates economic structure of the region to provide economic projections necessary for the appraisal of future needs.

ECONOMIC DEMAND SCHEDULE--The schedule of quantities of goods, services, or resources that will be purchased at various prices.

ECONOMIC EFFICIENCY--The situation in which productive resources are so allocated among alternative uses that any reshuffling from the pattern cannot improve any individual's position and still leave all the other individuals as well off as before.

ECONOMIC EVALUATION--The process through which the likely monetary benefits are compared to the monetary cost of a specific activity.

ECONOMIC JUSTIFICATION--A study to test the following items: (1) project benefits exceed project costs; (2) each separable segment or purpose provides benefits at least equal to its costs. Net benefits are maximized when the benefit of each segment or purpose is at least equal to the cost.

ECONOMIC LIFE--That period of time over which the project would serve a useful purpose; or the period of time after which further discounting of beneficial and adverse effects would have no appreciable impact.

ECONOMIC SURVEY--An analysis to determine the economic base.

ECOSYSTEM--The interacting system of a biological community and its environment.

EFFECTIVE PRECIPITATION--That portion of precipitation which remains on the foliage or in the soil that is available for evapotranspiration, and reduces the withdrawal of soil water by a like amount.

EFFECTIVENESS TEST--A measure of an alternative's technical performance in terms of how well it satisfies identified problems and meets projected needs.

EFFICIENCY TEST--Measurement and evaluation of all monetary and non-monetary effects, in order to identify from a monetary and non-monetary standpoint the least costly means of satisfying the identified problems and meeting the projected needs.

EVAPORATION--The physical process by which a liquid or solid is transformed to the gaseous state which in irrigation usually is restricted to the change of water from liquid to gas.

126

EVAPOTRANSPIRATION--The combined processes by which water is transferred from the earth surface to the atmosphere; evaporation of liquid or solid water plus transpiration from plants.

FEASIBILITY STUDY--A study to determine the engineering, economic and financial feasibility of a project and its environmental and social effects. Done in sufficient depth to give reasonable assurance to the project sponsors and financiers that the estimated costs will not be exceeded, that the benefits can be attained within the time and financial budgets presented, and the environmental and social effects will be as forecast.

FINANCIAL FEASIBILITY--A demonstration that beneficiaries are ready, willing, and able to pay reimbursable costs for products and services within the prescribed repayment period; that sufficient capital is authorized and available to finance construction to completion; and that estimated revenue to be derived during the prescribed repayment period is sufficient to cover reimbursable project costs.

FIRST COST--The total project construction cost including real estate, engineering, design, administration and supervision.

FIXED COSTS--Costs which are largely determined in advance of the year's operation and subject to little or no control on the part of the farmer. Rent of land, payment of taxes, interest on borrowed money and upkeep of buildings, fences and drains are examples of fixed costs.

FRAMEWORK PLAN--A broad guide to the best use, or combination of uses of water and related land resources of a region or subregion to meet foreseeable short- and long-term needs.

GROSS FARM INCOME--The total gross income realized by farm operators from farming. It includes cash receipts from the sale of farm products, government payments, value of food and fuel produced and consumed on farms where grown and rental value of farm dwellings.

HEADWATER BENEFITS--The downstream benefits realized by the storage or release of water by a reservoir project upstream. Application of the term is usually in reference to benefits to a downstream hydroelectric power plant.

HIGH LEVEL CROP YIELD--Harvested yield attained by the top five percent of the farmers using good management and related conservation practices and accounting for production hazards.

INCOME DISTRIBUTION--Refers to how the national income level is actually divided among individuals.

INCREMENTAL BENEFITS--The revenue generated by producing an extra batch of a product.

INDUCED COSTS--All uncompensated adverse effects caused by the construction and operation of a program or project, whether tangible or intangible.

127

INFRASTRUCTURE--The underlying foundation or basic framework, as of a system or organization.

INPUT-OUTPUT MODEL--The statistical measurement of the economic inputs and the outputs of all industries taken together in an inter-dependent system of commodity flows.

INSTALLATION COSTS--The value of goods and services necessary for the establishment of the project, including initial project construction; land, easements, rights-of-way, and water rights; capital outlays to relocate facilities or prevent damages; and all other expenditures for investigations and surveys, and designing, planning, and constructing a project after its authorization (excludes interest during construction). Also called project first costs.

INSTALLATION PERIOD--The time needed for the installation of project measures. The period could be only a few months for small pro-jects but could extend over a period of years in the case of large projects that would be installed in phases.

INTANGIBLE BENEFITS--(Secondary Benefits) Those benefits which, although recognized as having real value in satisfying human needs or desires, are not fully measurable in monetary terms, or are incapable of such expression in a formal analysis.

INTANGIBLE DAMAGES--Items of loss or damage for which a market price is not available or of undefined magnitude. These include such items as loss of life, creation of health hazards, inconveniences to transportation, etc.

INTERNAL RATE OF RETURN--The discount (interest rate which equalizes the present values of project benefit and cost streams (flows) over the life of the project.

IRRIGATION--Man's application of water to lands for growing of crops.

IRRIGATION EFFICIENCY--The ratio of the volume of water required for a specific beneficial use as compared to the volume of water delivered for this purpose. It is commonly interpreted as the volume of water stored in the soil for evapotranspiration compared to the volume of water delivered for this purpose, but may be defined and used in different ways.

IRRIGATION WATER REQUIREMENTS--Is the quanity of water exclusive of precipitation that is required for various beneficial uses.

JOINT COSTS--Costs of facilities that serve more than one project purpose.

JUSTIFICATION--The analysis of the costs and benefits (monetary and nonmonetary) of a project or program to determine whether or not the project or program should be implemented. See "economic justification."

LAND CAPABILITY CLASSIFICATION--Interpretive grouping of land primarily for agricultural purposes. Arable and nonarable soils are grouped according to their potentialities and limitations for sustained production of the commonly cultivated crops, or permanent vegetation, and according to the risk of soil damage.

LAND ENHANCEMENT BENEFITS--Those benefits resulting from the improved use of land made possible by a project.

LAND TREATMENT MEASURE--A tillage practice, or pattern, or land use, or land or management facility improvements to alter runoff, reduce sediment production, improve drainage and irrigation or improve plant or animal production.

LAND USE SURVEY--A quantitative determination of what land is being used for, usually done with aerial photography and field checking to determine type of land use, followed by measurement of the areas involved for the various uses.

LEACHING--The removal of soluble materials by the passage of water through soil.

LEACHING REQUIREMENT--The fraction of water entering the soil that must pass through the root zone in order to prevent soil salinity from exceeding a specific value.

MAJOR REPLACEMENT COSTS--Costs of replacement or rehabilitation of major structural or equipment items within the project of life.

MARKET DEMAND--An expression of the quantity of a product or service that can be sold at given prices.

MARKET VALUE--A price at which both buyers and sellers are willing to do business.

MITIGATION--Providing of services or facilities to compensate for project induced damages, usually related to fish and wildlife.

MULTIPLE PURPOSE PROJECT--A project with more than one purpose, e.g., a reservoir project which provides combinations of flood control, water supply (irrigation, municipal, industrial), hydroelectric power, recreation, wildlife and water quality improvements.

MULTIPLE USE (BEST)--The conscientious management of the various renewable resources such as water, wood, forage, wildlife, and recreation resources, to obtain sustained yield of products and services in the combination that will best meet the needs of the public now and in the future.

NATIONAL ECONOMIC DEVELOPMENT (NED)--Contributions to national economic development are increases in the net value of the national output of goods and services, expressed in monetary units. Contributions to NED are the direct benefits and costs that accrue in the planning area and the rest of the nation. Contributions to NED include increases in the net value of those goods and services that are marketed, and also of those that may not be marketed, such as outdoor recreation.

NEED--The difference between demands for a specified time and the capability of the existing level of resource development projected to the time considered.

CURRENT NEEDS--Current demands which are not being satisfied by the present level of resource development.

FUTURE NEEDS--The projected demands which are not satisfied by projected capability of the present level of development.

NEGATIVE BENEFITS--Conditions, brought about by a program or the construction or operation of a project, for which corrections require the expenditure of cost or effort which would not have been required had the project not been constructed, frequently called induced costs.

NET FARM INCOME--Net difference between gross returns from the sale of crops and the cost of production.

NET PRESENT VALUE (OR WORTH)--The summation of present values of the stream of project benefits less the present values of the stream of project costs, calculated at a specific discount (interest) rate.

NONREIMBURSABLE COSTS--Costs for which the financing agency does not seek repayment.

OPERATION, MAINTENANCE, AND REPLACEMENT COSTS--The value of goods and services needed to operate a constructed project and make repairs and replacements necessary to maintain the project in sound operating condition during its economic life or evaluation period.

OPPORTUNITY COST OF CAPITAL--The lowest acceptable rate of return which capital should be expected to earn in a given country.

OPTIMUM DEVELOPMENT--The optimum development of an area or a resource is that combination of scope and type of development which, when measured by economic, social, and other factors, best achieves the objectives of the development.

OPTIMUM PLANT GROWTH--The maximum rate of maturing, considering the type of plant, soil, water, fertilizer and cultural practices.

PAYMENT CAPACITY--The maximum ability of the bulk of agricultural water users in a specific area to pay annual average costs for water at their farm headgate on a per unit basis over a specified repayment period.

PER CAPITA WATER USE--The quantity per capita of water supplied in a municipality or district for a variety of uses or purposes during a given time period. It is usually taken to mean all uses included within the term "municipal use of water" and quantity wasted, lost, or otherwise unaccounted for.

PHYSIOGRAPHIC UNIT--An area in the landscape that internally has general similarity in selected range of environmental, topographic, and physical soil characteristics.

POINT ANALYSIS (ECONOMICS)--An analytical system which portrays an area's economy at a given point in time relative to some larger area (e.g., the nation) of which it is a part.

PRESENT VALUE (OR WORTH)--The value now of a sum, or sums, of money expended or received in the future. The present value varies in accordance with the discount (interest) rate selected.

PRICE ELASTICITY--The concept that a one percent change in the price received for a product results in a one percent change in the consumption of that product.

PRIMARY BENEFITS--(Direct Benefits) The value of goods and services directly resulting from the project, less associated costs incurred in realization of the benefits and any induced costs not included in project costs. Types of primary benefits may include domestic, municipal, and industrial water supply, irrigation, flood prevention, land stabilization, drainage, recreation, and fish and wildlife.

PROJECT--Any separable physical unit or closely related units, existing, undertaken or to be undertaken within a specific area for control and development of water and related land resources, which can be established and utilized independently or as an addition to an existing project, and can be, or has been, considered as a separate entity for purposes of evaluation.

PROJECT ECONOMIC COSTS--The value of all goods and services (land, labor, and materials) used in constructing, operating, and maintaining a project or program; interest during construction; and all other identifiable expenses, losses, liabilities, and induced adverse effects connected therewith, whether in goods or services, whether tangible or intangible, and whether or not compensation is involved. Project economic costs are the sum of installation costs; operation, maintenance, and replacement costs; and induced costs.

PROJECT FORMULATION--The process of establishing the components and nature of projects designed to attain some development goals or solve one or more problems. It involves a series of steps starting with determination of objectives by the decision makers, identification and definition of problems and needs, evaluation of available resources, development of alternative means of resolving problems and meeting the needs, evaluation of the alternatives and selection and implementation of the recommended plan. It is an orderly and systematic process which permits the interested public and decision makers to become aware of the assumptions made, data used, rationale and methodology employed, alternatives considered, cost, benefits, impacts and consequences of the alternatives and throughout the process to play a role in the decision making process.

RECREATION USE--Number of people using a recreation area--usually measured in recreation days, a visit by one individual to a recreational area for recreational purposes during any reasonable portion or all of a 24-hour period.

REPAYMENT PERIOD--The period of time over which the reimbursable costs of a project are repaid, generally 50 years.

ROOT ZONE--That part of the soil which is invaded by roots of plants. It is part of the zone of aeration which consists of soil and other materials that lie sufficiently near to the surface to discharge water into the atmosphere by the transpiration of plants or by evaporation from the soil.

SALINE SOIL--A nonalkali soil containing soluble salts in such quantities that they interfere with the growth of most plants.

SALINIZATION--The process by which salts accumulate in the soil.

SERVICE AREA--Territory in which a system is required or has the right to supply water, electric, or other services to customers.

SOCIO-CULTURAL IMPACTS--Impact (also known as "effects") are the economic and social consequences expected to result from alternative plans. The impacts of a plan are the measured changes between with the plan and the "without" conditions.

SOIL MOISTURE--Water in soil, divided into available and unavailable moisture. The former being water easily abstracted by roots of plants, while the latter is water held so firmly by adhesion and cohesive forces in the soil that it cannot usually be absorbed by plants rapidly enough to produce growth.

TRANSPIRATION--The process by which water in plants is transferred as water vapor to the atmosphere.

WATER CONSERVATION--A planned management of water to prevent waste or neglect.

WATER LOGGING--Filled or saturated with water or so filled with water as to be heavy or unmanageable, usually associated with soils.